Lecture Notes in Mathematics 2250

More information about this series at http://www.springer.com/series/304

Anna Skripka • Anna Tomskova

Multilinear Operator Integrals

Theory and Applications

 Springer

Anna Skripka
Department of Mathematics and Statistics
University of New Mexico
Albuquerque, NM, USA

Anna Tomskova
School of Computer Science and
Engineering
Inha University in Tashkent
Tashkent, Uzbekistan

ISSN 0075-8434   ISSN 1617-9692   (electronic)
Lecture Notes in Mathematics
ISBN 978-3-030-32405-6   ISBN 978-3-030-32406-3   (eBook)
https://doi.org/10.1007/978-3-030-32406-3

Mathematics Subject Classification (2010): Primary: 46L51, 47B49, 47A60, 47A63, 47B10, 47C15, 47A55, 15A60; Secondary: 46N50, 58J30, 46L87, 46G12, 46H10, 47L25, 26A16

This Springer imprint is published by the registered company Springer Nature Switzerland AG.
The registered company address is: Gewerbestrasse 11, 6330 Cham, Switzerland

Preface

A multilinear operator integral is a powerful tool in noncommutative analysis and its applications. Theory underlying multilinear operator integration has been developing since the 1950s, with a number of amazing advancements made in recent years. The field has accumulated many deep theoretical results and important applications, but no book on this beautiful and important subject appeared in the literature. This book provides a brief yet comprehensive treatment of multilinear operator integral techniques and their applications, partially filling the gap in the literature. The exposition is structured to be suitable for both a topics course and a research aid on methods, results, and applications of multilinear operator integrals.

We survey on earlier ideas and contributions to the field and then present in greater detail the best up-to-date results and modern methods. The content includes most practical, refined constructions of multiple operator integrals and fundamental technical results along with major applications of this tool to smoothness properties of operator functions (Lipschitz continuity, Hölder continuity, differentiability), approximation of operator functions, spectral shift functions, spectral flow in the setting of noncommutative geometry, quantum differentiability, and differentiability of noncommutative L^p norms. We demonstrate ideas and include proofs in simpler cases, while highly technical proofs are outlined and supplemented with a list of references. We also state selected open problems in the field.

Albuquerque, NM, USA   Anna Skripka
Tashkent, Uzbekistan   Anna Tomskova
September 2019

Acknowledgements

The authors thank Fedor Sukochev for inspiration to write an overview of multilinear operator integration, which has ultimately grown into this book. The authors are also grateful to the three referees for their valuable comments and suggestions. In particular, Theorems 5.1.12 and 5.1.13, Sects. 5.1.6 and 5.3.7, and the example after Theorem 3.3.11 were suggested by the referees.

Research of the first author was supported in part by NSF grant DMS-1554456.

Contents

Chapter 1
Introduction

A multilinear (or multiple) operator integral is a beautiful and powerful tool and one of the main objects of research in noncommutative analysis. In this book we discuss the rich theory of multiple operator integration and its numerous applications, choosing the latter to be our main motivation determining the exposition of the theory.

The first double operator integral technique can be detected in K. Löwner's work on matrix monotone functions in 1934, although the term "operator integral" was coined two decades later. K. Löwner related sign-definiteness of the increment of the matrix function $f(A) - f(B)$ to the sign-definiteness of the matrix $(f^{[1]}(\lambda_j, \mu_k))_{j,k=1}^d$ of the divided difference of f at the spectral points of self-adjoint matrices A and B by interpreting $f(A) - f(B)$ as the Schur product of the matrix $(f^{[1]}(\lambda_j, \mu_k))_{j,k=1}^d$ and the matrix $A - B$ in the eigen bases of A and B. The latter connection can be stated as

$$f(A) - f(B) = T_{f^{[1]}}^{A,B}(A - B), \tag{1.0.1}$$

where the linear transformation $T_{f^{[1]}}^{A,B}$ is an example of a double operator integral on the space of matrices. It is constructed based on the function $f^{[1]}$ of two variables and the spectral data of the matrices A and B. Later the representation (1.0.1) was extended to a much broader setting and found important applications in perturbation theory. In particular, it is utilized in the study of smoothness properties of operator functions and in derivation of various norm bounds for operator functions.

The first transformation receiving the name "double operator integral" was introduced by Yu. L. Daletskii and S. G. Krein in their work on differentiation of functions of infinite dimensional operators in 1956. They represented the derivative $\frac{d}{dt} f(A + tX)\big|_{t=0}$ of a function f along a path of bounded self-adjoint operators $A + tX$ as

$$\frac{d}{dt} f(A + tX)\big|_{t=0} = T_{f^{[1]}}^{A,A}(X),$$

© Springer Nature Switzerland AG 2019
A. Skripka, A. Tomskova, *Multilinear Operator Integrals*,
Lecture Notes in Mathematics 2250, https://doi.org/10.1007/978-3-030-32406-3_1

where the transformation $T_{f^{[1]}}^{A,A}$ is an iterated Riemann-Stieltjes integral with respect to the spectral family of A and the class of functions f in their approach is assumed to be more restrictive than turned out to be necessary. They also obtained an estimate for the operator norm of the derivative $\frac{d}{dt} f(A + tX)\big|_{t=0}$ based on the analysis of the transformation $T_{f^{[1]}}^{A,A}$. That estimate has been substantially improved and the restriction of f relaxed by means of modern approaches to double operator integration, but the idea of involving operator integrals in the study of differentiation and approximation of operator functions remains vital.

In 1964, M. G. Krein noticed a connection between the aforementioned results by K. Löwner and Yu. L. Daletski, S. G. Krein and brought it to the attention of B. S. Birman and M. Z. Solomyak, who later developed several approaches to double operator integration and substantially extended the range of applicability of the method. In particular, they introduced a transformation $T_\varphi^{A,B}$ acting on the Hilbert-Schmidt ideal $\mathcal{S}^2$ as an integral of a bounded measurable function φ with respect to a product spectral measure arising from the spectral measures of self-adjoint operators A and B. This transformation $T_\varphi^{A,B}$ is a double operator integral in the full sense due to the Hilbert space structure of the space $\mathcal{S}^2$. M. S. Birman and M. Z. Solomyak also introduced a transformation $T_\varphi^{A,B}$ on $\mathcal{B}(\mathcal{H})$, the space of bounded linear operators on a Hilbert space $\mathcal{H}$, that is tied to a factorization of the function $\varphi(\lambda, \mu)$ separating the variables λ and μ. The boundedness of $T_\varphi^{A,B}$ on $\mathcal{B}(\mathcal{H})$ and on the trace class $\mathcal{S}^1$ was characterized via the integral projective tensor product decomposition of φ by V. V. Peller in 1985. The latter result is similar to the celebrated Grothendieck's characterization of Schur multipliers on $\mathcal{B}(\ell^2)$ (see, e.g., [153, Chapter 5]).

While the aforementioned constructions of operator integrals along with the perturbation formula (1.0.1) allowed to obtain information about $f(A) - f(B)$, they were insufficient to characterize operator Lipschitz functions on Schatten-von Neumann ideals $\mathcal{S}^p$, $1 < p < \infty$. The latter was M. G. Krein's question stimulating the development of operator integration since 1964 and resolved by D. Potapov and F. Sukochev in 2011. The resolution was achieved by switching from the separation of variables in the symbol $\varphi = f^{[1]}$ of a double operator integral to a different decomposition of $f^{[1]}$ and subsequently applying harmonic analysis of UMD spaces. This approach also gave the best possible bound for the Schatten norm of $f(A) - f(B)$ in terms of $A - B$, namely,

$$\|f(A) - f(B)\|_p \leqslant c_p \|f\|_{\text{Lip}} \|A - B\|_p, \quad 1 < p < \infty, \tag{1.0.2}$$

and applied to all scalar Lipschitz functions f. Various counterexamples show that (1.0.2) does not extend to $p = 1$. The existing results in the case $A - B \in \mathcal{S}^1$ either involve a smaller set of admissible scalar functions f and a larger norm of $f^{[1]}$ or they estimate $f(A) - f(B)$ in the larger ideal $\mathcal{S}^{1,\infty}$.

As mentioned above, there are several constructions known under the name "double operator integral", which are denoted by $T_\varphi^{A,B}$ indicating that the function

φ, called a symbol, substitutes for an integrand and the spectral measures of the operators A and B substitute for a measure of integration. The properties of the transformation $T_\varphi^{A,B}$ depend on the space where it acts, on type of the symbol φ and sometimes on the operators A and B. These parameters and, hence, the specific construction to be used, are frequently stipulated by the class of problems approached with methods of operator integration. Often advantages of different approaches to operator integration should be synthesized for a successful resolution of a problem. In addition to double operator integrals on $\mathcal{B}(\mathcal{H})$ and S^p, there exist double operator integrals on $\mathcal{B}(X)$, where X is a Banach space, and on the noncommutative L^p-spaces. The latter are useful in the quantized calculus introduced by A. Connes and analytic approach to the Phillips-Atiyah-Patodi-Singer spectral flow. We compare various definitions of double operator integrals in Chap. 3 and discuss problems where they are utilized in Chap. 5.

Multilinear operator integrals naturally replace double operator integrals in higher order perturbation problems. For instance, for self-adjoint matrices A, B and n times differentiable functions f, we have the following extension of the representation (1.0.1):

$$f(A)-f(B)-\sum_{k=1}^{n-1}\frac{1}{k!}\frac{d^k}{dt^k}f(B+t(A-B))\Big|_{t=0} = T_{f^{[n]}}^{A,B,\dots,B}(A-B,\dots,A-B),$$

(1.0.3)

where $f^{[n]}$ is the nth order divided difference of f and $T_{f^{[n]}}^{A,B,\dots,B}$ is a n-linear Schur multiplier considered in the eigen bases of the matrices A and B. On the way to the representation (1.0.3), one also derives the formula

$$\frac{1}{k!}\frac{d^k}{dt^k}f(B+t(A-B))\Big|_{t=0} = T_{f^{[k]}}^{B,B,\dots,B}(A-B,\dots,A-B). \qquad (1.0.4)$$

The relation (1.0.3) suggests that in order to approximate a perturbed matrix function by a noncommutative analog of a Taylor polynomial with prescribed accuracy it suffices to find a suitable bound for the transformation $T_{f^{[n]}}^{A,B,\dots,B}$, which is an example of a multilinear operator integral on the product of matrix spaces.

The necessity to consider (1.0.3) for infinite dimensional operators A, B arose in mathematical physics when L. S. Koplienko attempted to extend the fundamental I. M. Lifshits–M. G. Krein trace formula

$$\mathrm{Tr}\,(f(A)-f(B)) = \int_{\mathbb{R}} f'(t)\xi_{A,B}(t)\,dt, \qquad (1.0.5)$$

where $\xi_{A,B} \in L^1(\mathbb{R})$ is the spectral shift function, to nontrace class perturbations $A-B$, for which the trace of $f(A)-f(B)$ is undefined, but the trace of (1.0.3) or its modifications can possibly be defined. Using double operator integrals he succeeded to prove an analog of (1.0.5) for Hilbert-Schmidt perturbations $A-B$ in 1984, but including more general perturbations required a development of a comprehensive theory of multilinear operator integration. Major steps of the development are briefly summarized below and discussed in a greater detail in Chap. 4, while applications of multilinear operator integration, including the trace formula (1.0.5) and its extensions, are discussed in Chap. 5.

As in the case of the double operator integrals, there exist several constructions known under the name "multiple operator integral", each introduced in response to a specific type of problems to be treated. The Birman–Solomyak double operator integral on $\mathcal{S}^2$ was extended by B. S. Pavlov to a multilinear transformation on the product space $\mathcal{S}^2 \times \cdots \times \mathcal{S}^2$ in 1969, and a recent approach to the same transformation by C. Coine, C. Le Merdy, F. Sukochev opened new technical opportunities. The transformation tied to a separation of variables of the symbol φ was extended from $\mathcal{B}(\mathcal{H})$ to $\mathcal{B}(\mathcal{H}) \times \cdots \times \mathcal{B}(\mathcal{H})$ by V. V. Peller in 2006 and to $\mathcal{I} \times \cdots \times \mathcal{I}$, where $\mathcal{I}$ is a symmetrically normed ideal of a semifinite von Neumann algebra satisfying property (F), by N. A. Azamov, A. L. Carey, P. G. Dodds, F. A. Sukochev independently in 2009. The latter extension allowed to prove existence of higher order derivatives for a broad set of functions f in a general setting and justify that the trace of (1.0.4) is well defined for $A - B \in \mathcal{S}^n$, but it did not capture nice properties of UMD ideals $\mathcal{S}^p$, $1 < p < \infty$, enabling sharper bounds that were essential for extending (1.0.5) to higher order Taylor remainders.

L. S. Koplienko's attempt to obtain an analog of (1.0.5) for $A - B \in \mathcal{S}^n$ was implemented in 2013 by D. Potapov, A. Skripka, F. Sukochev, who took an approach to multilinear operator transformations $T_{f^{[n]}}^{A,\dots,A}$ that does not involve separation of variables of the symbol. To derive bounds for the transformation on $\mathcal{S}^{p_1} \times \cdots \times \mathcal{S}^{p_n}$ with values in $\mathcal{S}^p$, where $1 < p, p_j < \infty$, $j = 1, \dots, n$, and $0 < \frac{1}{p} = \frac{1}{p_1} + \cdots + \frac{1}{p_n} < 1$, they used an intricate recursive procedure preserving the symbol $f^{[n]}$ modulo change of polynomial factors in the simplex integral representation for the divided difference and, thus, keeping the smallest norm of $f^{[n]}$. Apart from an estimate for $T_{f^{[n]}}^{B,\dots,B} : \mathcal{S}^{p_1} \times \cdots \times \mathcal{S}^{p_n} \to \mathcal{S}^p$ with p_j, p as above, they obtained the bound for the trace of the transformation

$$\left| \mathrm{Tr}\left(T_{f^{[n]}}^{B,\dots,B}(X, \dots, X) \right) \right| \leqslant c_n \| f^{(n)} \|_\infty \| X \|_n^n, \quad X \in \mathcal{S}^n, \tag{1.0.6}$$

leading to similar bounds for traces of higher order Taylor remainders and their representations in terms of higher order spectral shift functions.

One of main objectives in the study of multiple operator integrals consists in obtaining useful estimates for their norms. The existing variety of different multiple operator integral constructions is a by-product of the search for sharp estimates that were necessary to solve particular problems. Thus, a considerable part of this book is dedicated to discussion of various bounds for multiple operator integrals. In particular, we discuss the bounds that are derived based on one of the following methods: spectral integral representation of the transformation (Sects. 3.2 and 4.2), Hilbert space factorization of symbols (Sects. 3.3.3 and 4.3.1), reduction of order of polynomial integral momenta (Sects. 3.3.4 and 4.3.2), transference to noncommutative Calderón-Zygmund operators (Sect. 3.3.7). The spectral integral approach supplies the best bounds, but is limited to the Hilbert input space. The approach based on factorization of the symbol is fairly universal, but inevitably leads to a larger norm of the symbol. The order reduction of polynomial integral momenta provides the best bounds when the input space enjoys the UMD property. The transference approach allows to obtain the smallest norm of the symbol for a non-UMD space $\mathcal{S}^1$, but enlarges the output space to $\mathcal{S}^{1,\infty}$.

From the very beginning, development of multilinear operator integration has been motivated and directed by applications. The former has supplied an indispensable part of a toolkit in perturbation theory and quantised calculus. In Chap. 5, we discuss major applications of multilinear operator integration techniques and present a number of famous problems in operator theory, functional analysis, mathematical physics, and noncommutative geometry whose resolution relied on these techniques. The best up to date results on Lipschitz and Hölder estimates for operator functions, differentiation and approximation of operator functions, trace formulas and spectral shift functions, spectral flow in the von Neumann algebra setting, quantum differentiability, differentiability of noncommutative L^p-norms are presented along with main ideas or outlines of proofs.

Our exposition unfolds in order of increasing complexity. We start each section with discussion of the finite dimensional case that provides a basic insight into multiple operator integration without technicalities. We discuss properties and applications of finite dimensional multiple operator integrals or, equivalently, multilinear Schur multipliers independently of the general infinite dimensional theory. The connection between operator integrals and Schur multipliers discussed in this book allows to interchange results of both theories. We define continuous infinite dimensional versions of multiple operator integrals firstly on the Hilbert-Schmidt class, whose Hilbert space structure allows to derive the best results, secondly on Schatten-von Neumann ideals, where the UMD property, if holds, supplies second best results, and the algebra of bounded linear operators, next on noncommutative L^p-spaces and, finally, on general Banach spaces. We implement this program for double operator integrals in Chap. 3 and for general multiple operator integrals in Chap. 4. Thus, the reader looking only for multiple operator integrals in a specific setting can easily locate the respective material in the manuscript.

To master multiple operator integration in its full generality, the reader should be proficient in a graduate level functional analysis, harmonic analysis, general spectral theory of linear operators, basic theory of Schatten-von Neumann classes, noncommutative L^p-spaces, as well as symmetrically normed ideals and continuous traces. Nonetheless, the reader with background in linear algebra and basic analysis of functions and matrices can appreciate the theory of finite-dimensional multiple operator integrals (multilinear Schur multipliers); the reader familiar with general theory of linear operators on a Hilbert space can appreciate multiple operator integrals on $\mathcal{B}(\mathcal{H})$ and the Schatten-von Neumann ideals. To help the reader to recover the necessary background material, we recall major concepts and give references to relevant literature in Chap. 2.

Chapter 2
Notations and Preliminaries

In this chapter we recall major concepts from function spaces, theory of linear operators, ideals in von Neumann algebras and continuous traces, noncommutative L^p-spaces, Banach space theory, and approximation theory that are involved in our discussion of multiple operator integration and its applications. We also supply references to a systematic treatment of these concepts.

2.1 Spaces of Functions

We use the symbol $f : D \to R$ to denote a function f defined on a set D with values in a set R.

Let I be an interval in the real line $\mathbb{R}$ and let $f : I \to \mathbb{C}$, where $\mathbb{C}$ is the complex plane. The Lipschitz seminorm of f is defined by

$$\|f\|_{\mathrm{Lip}(I)} := \sup_{\substack{t,s \in I \\ t \neq s}} \frac{|f(t) - f(s)|}{|t - s|}.$$

We say that f is Lipschitz if and only if $\|f\|_{\mathrm{Lip}(I)} < \infty$. The class of all Lipschitz functions on I is denoted by $\mathrm{Lip}(I)$.

Let $\Lambda_\alpha(\mathbb{R})$ denote the set of Hölder functions of exponent $0 < \alpha < 1$, that is,

$$\Lambda_\alpha(\mathbb{R}) = \left\{ f : \mathbb{R} \to \mathbb{C} : \|f\|_{\Lambda_\alpha} := \sup_{\substack{t_1,t_2 \in \mathbb{R} \\ t_1 \neq t_2}} \frac{|f(t_1) - f(t_2)|}{|t_1 - t_2|^\alpha} < \infty \right\}.$$

In particular, $\Lambda_1(\mathbb{R}) = \mathrm{Lip}(\mathbb{R})$.

For $I \subset \mathbb{C}$, the linear space of continuous complex-valued functions on I is denoted by $C(I)$ and its subspace of bounded functions by $C_b(I)$. If $I \subset \mathbb{R}$, the

© Springer Nature Switzerland AG 2019

A. Skripka, A. Tomskova, *Multilinear Operator Integrals*,
Lecture Notes in Mathematics 2250, https://doi.org/10.1007/978-3-030-32406-3_2

space of functions continuously differentiable n times on I is denoted by $C^n(I)$ and its subspace of compactly supported functions by $C^n_c(I)$. The space of continuous functions on $\mathbb{R}$ that decay at infinity is denoted by $C_0(\mathbb{R})$.

For $\varphi \in C(\mathbb{T})$, where $\mathbb{T}$ is the unit circle, by its derivative at $z_0 \in \mathbb{T}$, we understand the limit

$$\varphi'(z_0) := \lim_{\mathbb{T}\ni z \to z_0} \frac{\varphi(z) - \varphi(z_0)}{z - z_0}, \tag{2.1.1}$$

provided it exists. The symbol $C^n(\mathbb{T})$ denotes the set of functions n times continuously differentiable on $\mathbb{T}$ in the sense of (2.1.1).

Let Ω be a subset in $\mathbb{R}^n$ endowed with a Borel σ-algebra and measure μ. Let $L^\infty(\Omega, \mu)$ denote the space of all complex-valued, essentially bounded functions on (Ω, μ). The ess sup norm on $L^\infty(\Omega, \mu)$ is denoted $\| \cdot \|_\infty$. When μ is the Lebesgue measure we write $L^\infty(\Omega)$ instead of $L^\infty(\Omega, \mu)$. For $1 \leqslant p < \infty$, let $L^p(\Omega, \mu)$ denote the space of all measurable functions $f : \Omega \to \mathbb{C}$ satisfying $\|f\|_p := \left(\int_\Omega |f|^p d\mu\right)^{1/p} < \infty$. The space $L^p(\Omega, \mu)$ is equipped with the norm $\| \cdot \|_p$. When $\Omega = \mathbb{N}$, where $\mathbb{N}$ is the set of positive integers, and μ is a counting measure, $L^p(\Omega, \mu)$ becomes the space of p-summable sequences, which we denote ℓ^p.

Let $\mathcal{F}f$ and $\mathcal{F}^{-1}f$ be the Fourier transform and the inverse Fourier transform, respectively, of the function $f \in L^1(\mathbb{R})$, that is,

$$\mathcal{F}f(t) = \frac{1}{\sqrt{2\pi}} \int_\mathbb{R} f(s)\,e^{-ist}\,ds, \quad \mathcal{F}^{-1}f(s) = \frac{1}{\sqrt{2\pi}} \int_\mathbb{R} f(t)\,e^{ist}\,dt.$$

Given $n \in \mathbb{N}$, denote the Wiener space

$$W_n(\mathbb{R}) = \{f \in C^n(\mathbb{R}) : f^{(k)}, \mathcal{F}f^{(k)} \in L^1(\mathbb{R}),\ k = 0, \ldots, n\}.$$

By standard methods of harmonic analysis one can see that $C^{n+1}_c(\mathbb{R})$ is a dense subset of $W_n(\mathbb{R})$.

Let $\mathcal{R}$ denote the set of bounded rational complex-valued functions on $\mathbb{R}$ with non-real poles, $\mathcal{R}_+$ (respectively, $\mathcal{R}_-$) the subset of $\mathcal{R}$ consisting of functions with poles in the upper half-plane (respectively, in the lower half-plane).

Let $0 < p, q \leqslant \infty$, $s \in \mathbb{R}$, and let $\tilde{B}^s_{pq}$ denote a modified homogenous Besov space either on $\mathbb{R}$ or on $\mathbb{T}$. In the context of perturbation theory, we are particularly interested in the case $p = \infty$, $q = 1$, and $s \in \mathbb{N} \cup \{0\}$. Let $w_0 \in C^\infty(\mathbb{R})$ be such that its Fourier transform is supported in $[-2, -1/2] \cup [1/2, 2]$ and $\mathcal{F}w_0(y) + \mathcal{F}w_0(y/2) = 1$ for $1 \leqslant y \leqslant 2$ and define $w_k(x) = 2^k w_0(2^k x)$ for $x \in \mathbb{R}$, $k \in \mathbb{Z}$, where $\mathbb{Z}$ is the set of integers. Then, for $n \in \mathbb{N} \cup \{0\}$,

$$\tilde{B}^n_{\infty 1}(\mathbb{R}) = \left\{ f \in C^n(\mathbb{R}) : \|f^{(n)}\|_\infty + \sum_{k \in \mathbb{Z}} 2^{nk} \|f * w_k\|_\infty < \infty \right\}.$$

Alternatively, the Besov space can be characterized as follows:

$$f \in \tilde{B}_{\infty 1}^n(\mathbb{R}) \iff f(x) = c_0 + c_1 x + \cdots + c_n x^n + f_0(x),$$

$$c_j \in \mathbb{C}, \quad j = 0, \ldots, n, \quad \sum_{k \in \mathbb{Z}} 2^{nk} \|f_0 * w_k\|_\infty < \infty, \quad \operatorname{supp} \mathcal{F} f_0 \subset \mathbb{R} \setminus \{0\}.$$

The space $\tilde{B}_{\infty 1}^n(\mathbb{R})$ is sometimes considered with the seminorm

$$\|f\|_{\tilde{B}_{\infty 1}^n(\mathbb{R})} = \sum_{k \in \mathbb{Z}} 2^{nk} \|f * w_k\|_\infty$$

and sometimes with the seminorm

$$\|f\|_{B_{\infty 1}^n(\mathbb{R})} = \|f^{(n)}\|_\infty + \|f\|_{\tilde{B}_{\infty 1}^n(\mathbb{R})}.$$

When the Besov space is considered with the latter seminorm, we denote it $B_{\infty 1}^n(\mathbb{R})$. There are other deviations in definitions and notations of Besov spaces in different publications. In particular, initially Besov spaces $B_{\infty 1}^n(\mathbb{R})$ were defined via summability of iterated difference operators. For a detailed exposition of Besov spaces we refer the reader to [140, 197, 207].

Sometimes it is simpler to work with subclasses of $B_{\infty 1}^n(\mathbb{R})$. In particular,

$$W_n(\mathbb{R}) \subset B_{\infty 1}^n(\mathbb{R}).$$

Another known inclusion is

$$\{f \in C^{n+1}(\mathbb{R}) : f^{(n)}, f^{(n+1)} \in L^2(\mathbb{R})\} \subset B_{\infty 1}^n(\mathbb{R})$$

(see [157, Lemma 7]). It is also known that

$$\{f \in C^n(\mathbb{R}) : f^{(n-1)} \in \Lambda_{1-\epsilon}(\mathbb{R}), f^{(n)} \in \Lambda_\epsilon(\mathbb{R})\} \subset B_{\infty 1}^n(\mathbb{R})$$

for $0 < \epsilon < 1$ (see [157, Theorem 4 and Remark 5]).

For a function f on the unit circle $\mathbb{T}$ by $\mathcal{F} f$ we define its Fourier series with coefficients denoted by $\mathcal{F} f(n)$, $n \in \mathbb{Z}$. Let $\mathcal{A}(\mathbb{D})$ denote the disc algebra, that is, the algebra of functions $f : \bar{\mathbb{D}} \to \mathbb{C}$ analytic on the open unit disc $\mathbb{D}$ and continuous on the closed unit disc $\bar{\mathbb{D}}$. The algebra $\mathcal{A}(\mathbb{D})$ can be naturally identified with the algebra of continuous functions on the unit circle $\mathbb{T}$ with vanishing Fourier coefficients $\mathcal{F}(-n)$, $n \in \mathbb{N}$. Given $n \in \mathbb{N}$, denote

$$W_n(\mathbb{T}) = \{f : \mathbb{T} \to \mathbb{C} : f^{(k)} \in L^1(\mathbb{T}), \mathcal{F} f^{(k)} \in \ell^1(\mathbb{Z}), k = 0, \ldots, n\}.$$

Let $n \in \mathbb{N}$. The Besov class $B_{\infty 1}^n(\mathbb{T})$ of functions on $\mathbb{T}$ can be defined as follows. Let $w \in C^\infty(\mathbb{R})$ be such that

$$w \geqslant 0, \quad \operatorname{supp} w \subset \left[\frac{1}{2}, 2 \right], \quad \text{and} \quad w(x) = 1 - w\left(\frac{x}{2} \right) \quad \text{for } x \in [1, 2].$$

Consider the trigonometric polynomials w_m and $w_m^\sharp$ defined by

$$w_m(z) = \sum_{k \in \mathbb{Z}} w\left(\frac{k}{2^m} \right) z^k, \quad m \geqslant 1, \quad w_0(z) = \bar{z} + 1 + z, \quad \text{and}$$

$$w_m^\sharp(z) = \overline{w_m(z)}, \quad m \geqslant 0.$$

Then, for each function φ on $\mathbb{T}$,

$$\varphi = \sum_{m \geqslant 0} \varphi * w_m + \sum_{m \geqslant 1} \varphi * w_m^\sharp.$$

The Besov class $B_{\infty 1}^n(\mathbb{T})$ consists of functions φ on $\mathbb{T}$ such that

$$\{2^{nm} \|\varphi * w_m\|_\infty\}_{m \geqslant 0} \in \ell^1 \quad \text{and} \quad \{2^{nm} \|\varphi * w_m^\sharp\|_\infty\}_{m \geqslant 1} \in \ell^1$$

and is considered with the seminorm

$$\|\varphi\|_{B_{\infty 1}^n(\mathbb{T})} = \sum_{k \geqslant 0} 2^{nk} \|\varphi * w_k\|_\infty + \sum_{k \geqslant 1} 2^{nk} \|\varphi * w_k^\sharp\|_\infty.$$

We will frequently use the fact that for $n_1, n_2 \in \mathbb{N}$,

$$B_{\infty 1}^{n_2}(\mathbb{T}) \subset B_{\infty 1}^{n_1}(\mathbb{T}) \quad \text{if } n_1 < n_2.$$

We also have

$$W_n(\mathbb{T}) \subset B_{\infty 1}^n(\mathbb{T}).$$

2.2 Divided Differences

We recall that the divided difference of the zeroth order $f^{[0]}$ is the function f itself. Let $\lambda_0, \lambda_1, \ldots, \lambda_n$ be points in $\mathbb{R}$ (respectively, in $\mathbb{T}$) and let $f \in C^n(\mathbb{R})$ (respectively, $f \in C^n(\mathbb{T})$). The divided difference $f^{[n]}$ of order n is defined recursively by

$$f^{[n]}(\lambda_0, \ldots, \lambda_n) = \lim_{\lambda \to \lambda_n} \frac{f^{[n-1]}(\lambda_0, \ldots, \lambda_{n-2}, \lambda) - f^{[n-1]}(\lambda_0, \ldots, \lambda_{n-2}, \lambda_{n-1})}{\lambda - \lambda_{n-1}}.$$

$$(2.2.1)$$

Basic properties of the divided difference of a function defined on $\mathbb{R}$ can be found in, for example, [65, Section 4.7]. In particular, for $f \in C^n(\mathbb{R})$ and I an interval in $\mathbb{R}$,

$$f^{[n]}(\lambda_0, \ldots \lambda_0) = \frac{1}{n!} f^{(n)}(\lambda_0),$$

$$\left\| f^{[n]} \right\|_{L^\infty(I^{n+1})} \leqslant \frac{1}{n!} \left\| f^{(n)} \right\|_{L^\infty(I)}, \tag{2.2.2}$$

and the divided difference is invariant with respect to any permutation of its variables.

By [191, Lemma 3.2] (which follows directly from [59, Theorem 2.1 and Lemma 2.2]), for $f \in C^n(\mathbb{T})$, we have

$$\left\| f^{[n]} \right\|_{L^\infty(\mathbb{T}^{n+1})} \leqslant \frac{\pi^{(n+3)/2}}{2^{n+1}\Gamma((n+1)/2)} \left\| f^{(n)} \right\|_{L^\infty(\mathbb{T})}, \tag{2.2.3}$$

where $\Gamma(\cdot)$ is the Gamma function.

For $f \in \mathrm{Lip}(I)$ we have

$$\sup_{\substack{\lambda, \mu \in I \\ \lambda \neq \mu}} |f^{[1]}(\lambda, \mu)| \leqslant \|f\|_{\mathrm{Lip}(I)}. \tag{2.2.4}$$

2.3 Linear Operators

Let X be a Banach space equipped with the norm $\| \cdot \|_X$ and $\mathcal{Y}$ a Banach space equipped with the norm $\| \cdot \|_{\mathcal{Y}}$. Denote by $\mathcal{B}(X, \mathcal{Y})$ the Banach space of bounded linear operators mapping X to $\mathcal{Y}$ and equipped with the operator norm

$$\|X\| := \sup_{\|\xi\|_X \leqslant 1} \|X(\xi)\|_{\mathcal{Y}}, \quad X \in \mathcal{B}(X, \mathcal{Y}).$$

When we need to distinguish between different norms we use more detailed symbols $\| \cdot \|_{X \to \mathcal{Y}}$ or $\| \cdot : X \to \mathcal{Y} \|$ instead of $\| \cdot \|$. If $X = \mathcal{Y}$, we write $\mathcal{B}(X)$ instead of $\mathcal{B}(X, X)$.

We identify the algebraic tensor product $X^* \otimes \mathcal{Y}$ with the space of finite rank operators in $\mathcal{B}(X, \mathcal{Y})$ via

$$(x^* \otimes y)(x) := x^*(x)y, \quad x \in X, \ x^* \in X^*, \ y \in \mathcal{Y}.$$

We will also work with multilinear transformations between Banach spaces, that is, transformations that are linear in each of the variables separately. If $X_1 \ldots, X_n$

are Banach spaces, then the symbol $\mathcal{B}_n(\mathcal{X}_1 \times \cdots \times \mathcal{X}_n, \mathcal{Y})$ stands for the space of n-linear mappings $\mathcal{X}_1 \times \cdots \times \mathcal{X}_n \to \mathcal{Y}$ equipped with the norm

$$\|X\| := \sup_{\|\xi_1\|_{\mathcal{X}_1}, \ldots, \|\xi_n\|_{\mathcal{X}_n} \leqslant 1} \|X(\xi_1, \ldots, \xi_n)\|_{\mathcal{Y}}, \quad X \in \mathcal{B}_n(\mathcal{X}_1 \times \cdots \times \mathcal{X}_n, \mathcal{Y}).$$

Again we can also use $\| \cdot \|_{\mathcal{X}_1 \times \cdots \times \mathcal{X}_n \to \mathcal{Y}}$ or $\|\cdot : \mathcal{X}_1 \times \cdots \times \mathcal{X}_n \to \mathcal{Y}\|$ instead of $\| \cdot \|$. By $I_{\mathcal{X}} \in \mathcal{B}(\mathcal{X})$ we denote the identity operator on $\mathcal{X}$. When there is no ambiguity, we simply write I for the identity operator.

Let $\mathcal{H}$ be a separable Hilbert space equipped with the inner product $\langle \cdot, \cdot \rangle$ and the respective norm $\| \cdot \|$. We will consider both an infinite dimensional Hilbert space $\mathcal{H}$ and the d-dimensional space $\ell_d^2, d \in \mathbb{N}$. In the case $\mathcal{H} = \ell_d^2$, we consider the canonical inner product. Let $\mathcal{B}(\mathcal{H})$ denote the C^*-algebra of bounded linear operators on $\mathcal{H}$ equipped with the operator norm. The subset of the normal operators in $\mathcal{B}(\mathcal{H})$ is denoted by $\mathcal{B}_{norm}(\mathcal{H})$ and of the self-adjoint operators by $\mathcal{B}_{sa}(\mathcal{H})$. The set of closed self-adjoint operators defined on a dense subset D of $\mathcal{H}$ is denoted $\mathcal{D}_{sa}$. When we write $A, B \in \mathcal{D}_{sa}$, we assume that A, B are densely defined on the same subset of $\mathcal{H}$.

We will freely apply facts from basic spectral theory of self-adjoint and, more generally, normal operators that can be found in, for example, [38, Chapters 5 and 6]. Let P_ξ denote the orthogonal projection on the unit vector $\xi \in \mathcal{H}$, that is,

$$P_\xi(\cdot) = \langle \cdot, \xi \rangle \xi.$$

Let $\sigma(A)$ denote the spectrum of a linear operator A densely defined in $\mathcal{H}$. If $A \in \mathcal{D}_{sa}$, then $\sigma(A) \subset \mathbb{R}$. Let E_A denote the spectral measure of a normal operator A. Given a Borel function f, let $f(A)$ denote the operator function defined by the standard functional calculus.

We will work with the tensor product $A \otimes B$ of a $d \times d$-matrix A and $m \times m$-matrix B, which is the $dm \times dm$-matrix given by

$$A \otimes B = \begin{pmatrix} a_{11}B & \cdots & a_{1d}B \\ \vdots & & \vdots \\ a_{d1}B & \cdots & a_{dd}B \end{pmatrix}.$$

We will also touch upon tensor product spaces of infinite dimensional operators. For details on this account we refer the reader to [174, 202].

2.4   Schatten-von Neumann Classes

Results recalled in this section and a more detailed discussion of ideals of compact operators can be found in, for instance, [38, 85, 183].

Let an infinite dimensional separable Hilbert space $\mathcal{H}$ be fixed. Let S^∞ denote the Banach space of all compact operators on $\mathcal{H}$ equipped with the uniform norm. Recall that every $X \in S^\infty$ has a representation

$$X = \sum_{n=1}^{\infty} \lambda_n(X) P_n,$$

where $\{\lambda_n(X)\}_{n=1}^{\infty}$ is a sequence of eigenvalues of the operator X such that the sequence $\{|\lambda_n(X)|\}_{n=1}^{\infty}$ is decreasing (that is, not increasing) and tends to 0 and P_n, $n \in \mathbb{N}$, is a rank one projection on $\mathcal{H}$. Denote by $\mathfrak{F}$ the algebra of all finite rank operators on $\mathcal{H}$, by $\mathfrak{F}_{norm}$ its subset of normal and by $\mathfrak{F}_{sa}$ its subset of self-adjoint operators.

Let $1 \leqslant p < \infty$. By S^p we denote the pth Schatten(-von Neumann) ideal, that is, the Banach ideal of all $X \in S^\infty$ such that

$$\|X\|_p := (\mathrm{Tr}(|X|^p))^{1/p} < \infty,$$

where $|X| = (X^*X)^{1/2}$ and $\mathrm{Tr}(\cdot)$ is the standard trace extending the canonical matrix trace, that is,

$$\mathrm{Tr}(A) = \sum_{n=1}^{\infty} \lambda_n(A)$$

for a positive semidefinite operator A. When $\mathcal{H} = \ell_d^2$, we denote the respective Schatten-von Neumann ideal $(\mathcal{B}(\ell_d^2), \|\cdot\|_p)$ by S_d^p.

By $\|X\| = \|X\|_\infty$ we denote the operator norm of $X \in \mathcal{B}(\mathcal{H})$. The set of finite rank operators is dense in every S^p considered with the norm $\|\cdot\|_p$, $1 \leqslant p \leqslant \infty$.

Let $\{s_n(X)\}_{n=1}^{\infty}$ denote the decreasing sequence of singular values of X, that is, the decreasing sequence of eigenvalues of $|X|$. For $1 \leqslant p < \infty$, $X \in S^p$ if and only if $\{s_n(X)\}_{n=1}^{\infty}$ is an element of the Banach sequence space ℓ^p, that is, $\sum_{n=1}^{\infty} s_n(X)^p < \infty$. In particular, we have the inclusion

$$S^p \subsetneq S^q, \quad p < q,$$

and the inequalities

$$\|X\| \leqslant \|X\|_q \leqslant \|X\|_p.$$

The norm $\|\cdot\|_p$ satisfies the Hölder inequality

$$\|XY\|_r \leqslant \|X\|_p \|Y\|_q, \tag{2.4.1}$$

where $1 \leqslant p, q, r \leqslant \infty$ are such that $\frac{1}{r} = \frac{1}{p} + \frac{1}{q}$ and $X \in S^p$, $Y \in S^q$. We have the duality

$$(S^p)^* = S^q, \quad \frac{1}{p} + \frac{1}{q} = 1,$$

with every functional φ on S^p in the form $\varphi(X) = \mathrm{Tr}\,(XY_\varphi)$ for some $Y_\varphi \in S^q$, $\|\varphi\| = \|Y_\varphi\|_q$.

The ideal S^1 is called the trace class ideal and S^2 the Hilbert-Schmidt class. The ideal S^2 with the inner product

$$\langle X, Y \rangle := \mathrm{Tr}(Y^*X), \quad X, Y \in S^2,$$

and corresponding norm $\| \cdot \|_2$ becomes a Hilbert space. The Hilbert-Schmidt norm of a finite-dimensional operator $X \in \mathcal{B}(\ell_d^2)$ can also be computed via a matrix representation of $X = \left(x_{jk}\right)_{j,k=1}^d$ as follows:

$$\|X\|_2 = \Big(\sum_{j,k=1}^d |x_{jk}|^2 \Big)^{1/2}. \tag{2.4.2}$$

The weak Schatten(-von Neumann) ideal $S^{p,\infty}$, $1 \leqslant p < \infty$, is defined by

$$S^{p,\infty} = \Big\{ X \in S^\infty : \sup_{n \in \mathbb{N}} n^{\frac{1}{p}} s_n(X) < \infty \Big\}.$$

The space $S^{p,\infty}$ equipped with the quasi-norm

$$\|X\|'_{p,\infty} := \sup_{n \in \mathbb{N}} n^{\frac{1}{p}} s_n(X)$$

is a quasi-Banach space. If $p > 1$, there exists an equivalent norm

$$\|X\|_{p,\infty} := \sup_{N \in \mathbb{N}} N^{\frac{1}{p}-1} \sum_{k=1}^N s_k(X)$$

with respect to which $S^{p,\infty}$ is a Banach space. We have the inclusion

$$S^p \subset S^{p,\infty} \subset S^q$$

for $1 \leqslant p < q < \infty$.

2.5 Product of Spectral Measures

In this section we collect preliminaries on products of spectral measures crucial in the definition of a double operator integral.

Let (Ω_j, Σ_j), $j = 1, 2$ be a measurable space. Let E and F be spectral measures given on Σ_1 and Σ_2, respectively, with values in the set of orthogonal projections in $\mathcal{B}(\mathcal{H})$.

Let us consider

$$\mathcal{E}(\sigma_1) : X \mapsto E(\sigma_1)X, \quad \sigma_1 \in \Sigma_1, \ \sigma_2 \in \Sigma_2, \ X \in \mathcal{S}^2.$$
$$\mathcal{F}(\sigma_2) : X \mapsto XF(\sigma_2),$$

It is clear that $\mathcal{E}$ and $\mathcal{F}$ are commuting spectral measures on (Ω_1, Σ_1) and (Ω_2, Σ_2), respectively, with respect to $\mathcal{S}^2$. Define the product of the measures $\mathcal{E}$ and $\mathcal{F}$

$$G(\sigma_1 \times \sigma_2) := \mathcal{E}(\sigma_1)\mathcal{F}(\sigma_2), \quad \sigma_1 \in \Sigma_1, \ \sigma_2 \in \Sigma_2, \tag{2.5.1}$$

so that

$$G(\sigma_1 \times \sigma_2)(X) = E(\sigma_1)XF(\sigma_2), \quad X \in \mathcal{S}^2. \tag{2.5.2}$$

To verify that G is indeed a spectral measure, we need to show that G is σ-additive. We note that the product of two spectral measures can fail to be σ-additive for an arbitrary Hilbert space, as it is shown in [43]. In [40, Theorem 2] it is established that G constructed above is in fact σ-additive. The proof uses the specificity of the fact that G is defined with respect to $\mathcal{S}^2$, but not an arbitrary Hilbert space. A more general result for the product of n spectral measures, $n \geqslant 2$ is proved in [138, Theorem 1]. We present the proof from [40, Theorem 2] here for the reader's convenience.

Proposition 2.5.1 *Let (Ω_1, Σ_1) and (Ω_2, Σ_2) be measure spaces, $\mathcal{H}$ a separable Hilbert space and let E and F be spectral measures with respect to $\mathcal{H}$ on (Ω_1, Σ_1) and (Ω_2, Σ_2), respectively. Then, the mapping $G : \Sigma_1 \times \Sigma_2 \to \mathcal{B}(\mathcal{S}^2)$ defined by (2.5.2) extends to a spectral measure on (Ω, Σ) with respect to $\mathcal{S}^2$, where $\Omega = \Omega_1 \times \Omega_2$ and $\Sigma = \Sigma_1 \otimes \Sigma_2$ is the minimal σ-algebra generated by the algebra $\Sigma_1 \times \Sigma_2$ of "measurable rectangles".*

Proof If $\sigma \in \Sigma_1 \times \Sigma_2$ is such that $\sigma = \bigcup_{k=1}^{n} \sigma_1^{(k)} \times \sigma_2^{(k)}$, where $\left(\sigma_1^{(i)} \times \sigma_2^{(i)}\right) \cap \left(\sigma_1^{(j)} \times \sigma_2^{(j)}\right) = \varnothing$, for $i \neq j$, then we define

$$G(\sigma)X := \sum_{k=1}^{n} G(\sigma_1^{(k)} \times \sigma_2^{(k)})X. \tag{2.5.3}$$

Correctness of the definition above can be checked using standard arguments.

Let us prove that G takes values in orthogonal projections on $\mathcal{S}^2$. Indeed, for $X, Y \in \mathcal{S}^2$ and $\delta = \sigma_1 \times \sigma_2$, $\sigma_1 \in \Sigma_1$, $\sigma_2 \in \Sigma_2$, we have that $G(\delta)^2 = G(\delta)$ and

$$
\begin{aligned}
\langle G(\delta)X, Y \rangle &= \langle E(\sigma_1)XF(\sigma_2), Y \rangle = \mathrm{Tr}(Y^* E(\sigma_1)XF(\sigma_2)) \\
&= \mathrm{Tr}((E(\sigma_1)YF(\sigma_2))^* X) = \langle X, E(\sigma_1)YF(\sigma_2) \rangle = \langle X, G(\delta)Y \rangle,
\end{aligned}
$$

that is, $G(\delta)^* = G(\delta)$. For an arbitrary set $\delta \in \Sigma_1 \times \Sigma_2$ the proof is similar using (2.5.3). Since E and F are spectral measures, it follows that

$$
G(\Omega)X = E(\Omega_1)XF(\Omega_2) = X,
$$

that is, $G(\Omega)$ is the identity operator on $\mathcal{S}^2$.

We show that G is σ-additive on $\Sigma_1 \times \Sigma_2$. For $X, Y \in \mathcal{S}^2$ consider the scalar measure

$$
\nu_{X,Y} : \sigma_1 \times \sigma_2 \mapsto \langle G(\sigma_1 \times \sigma_2)X, Y \rangle, \quad \sigma_1 \in \Sigma_1, \ \sigma_2 \in \Sigma_2.
$$

Let $\delta \in \Sigma_1 \otimes \Sigma_2$ be such that $\delta = \bigcup_{n=1}^{\infty} \delta_n$, $\delta_n \in \Sigma_1 \times \Sigma_2$ and $\delta_n \cap \delta_m = \varnothing$, $n \neq m$, $n, m \in \mathbb{N}$. Recalling the inequality

$$
\sum_{n=1}^{\infty} \| G(\delta_n)X \|_2^2 \leqslant \| X \|_2^2, \quad X \in \mathcal{S}^2,
$$

(see, e.g., [123, Lemma 12.4.7]) and evaluating

$$
\begin{aligned}
\sum_{n=1}^{\infty} |\langle G(\delta_n)X, Y \rangle| &= \sum_{n=1}^{\infty} |\langle G(\delta_n)X, G(\delta_n)Y \rangle| \\
&\leqslant \sum_{n=1}^{\infty} \| G(\delta_n)X \|_2 \| G(\delta_n)Y \|_2 \\
&\leqslant \Big(\sum_{n=1}^{\infty} \| G(\delta_n)X \|_2^2 \Big)^{\frac{1}{2}} \Big(\sum_{n=1}^{\infty} \| G(\delta_n)Y \|_2^2 \Big)^{\frac{1}{2}} \\
&\leqslant \| X \|_2 \| Y \|_2,
\end{aligned}
\tag{2.5.4}
$$

we obtain that the series $\sum_{n=1}^{\infty} \langle G(\delta_n)X, Y \rangle$ converges for any $X, Y \in \mathcal{S}^2$.

Next we prove that $\nu_{X,Y}$ is a σ-additive scalar measure, that is,

$$
\Big\langle G\Big(\bigcup_{n=1}^{\infty} \delta_n \Big)X, Y \Big\rangle = \sum_{n=1}^{\infty} \langle G(\delta_n)X, Y \rangle.
\tag{2.5.5}
$$

If $X = \langle \cdot, \xi_2 \rangle \xi_1$ and $Y = \langle \cdot, \eta_2 \rangle \eta_1$ for some $\xi_j, \eta_j \in \mathcal{H}$, and $\sigma_j \in \Sigma_j$, $j = 1, 2$, then we have that

$$\nu_{X,Y^*}(\sigma_1 \times \sigma_2) = \mathrm{Tr}(YE(\sigma_1)XF(\sigma_2)) = \langle E(\sigma_1)\xi_1, \eta_2 \rangle \langle F(\sigma_2)\eta_1, \xi_2 \rangle,$$

which is a product of two scalar σ-additive measures. Thus, $\nu_{X,Y}$ is σ-additive for any one-dimensional operators X and Y. Therefore, any one-dimensional operators X and Y satisfy (2.5.5).

If each of X and Y is represented by a finite sum of one-dimensional operators, then by linearity of the inner product, we obtain that X and Y satisfy (2.5.5).

Let now $X, Y \in \mathcal{S}^2$ and $\{X_k\}_{k \geqslant 1}$ and $\{Y_k\}_{k \geqslant 1}$ be sequences of finite-dimensional operators such that $X_k \to X$ and $Y_k \to Y$ in $\mathcal{S}^2$ as $k \to \infty$. Since $\delta = \bigcup_{n=1}^{\infty} \delta_n \in \Sigma_1 \otimes \Sigma_2$, it follows from (2.5.3) that

$$\langle G(\delta)X_k, Y_k \rangle \to \langle G(\delta)X, Y \rangle \quad \text{as } k \to \infty. \tag{2.5.6}$$

Applying (2.5.4), we have that

$$\left| \sum_{n=1}^{\infty} \langle G(\delta_n)X_k, Y_k \rangle - \sum_{n=1}^{\infty} \langle G(\delta_n)X, Y \rangle \right|$$

$$\leqslant \sum_{n=1}^{\infty} \left| \langle G(\delta_n)X_k, Y_k \rangle - \langle G(\delta_n)X, Y \rangle \right|$$

$$\leqslant \sum_{n=1}^{\infty} \left| \langle G(\delta_n)X_k, Y_k \rangle - \langle G(\delta_n)X, Y_k \rangle \right| + \sum_{n=1}^{\infty} \left| \langle G(\delta_n)X, Y_k \rangle - \langle G(\delta_n)X, Y \rangle \right|$$

$$= \sum_{n=1}^{\infty} \left| \langle G(\delta_n)(X_k - X), Y_k \rangle \right| + \sum_{n=1}^{\infty} \left| \langle G(\delta_n)X, Y_k - Y \rangle \right|$$

$$\leqslant \|X_k - X\|_2 \|Y_k\|_2 + \|X\|_2 \|Y_k - Y\|_2.$$

Hence,

$$\sum_{n=1}^{\infty} \langle G(\delta_n)X_k, Y_k \rangle \to \sum_{n=1}^{\infty} \langle G(\delta_n)X, Y \rangle \quad \text{as } k \to \infty,$$

which along with (2.5.6) proves (2.5.5) for all $X, Y \in \mathcal{S}^2$. The observation that σ-additivity of $\nu_{X,Y}$ implies σ-additivity of G completes the proof. $\square$

2.6 Classical Noncommutative L^p-Spaces and Weak L^p-Spaces

In this section we recall definitions and some important properties of the classical noncommutative L^p-spaces and the weak L^p-spaces. Details can be found in, for instance, [123, 154, 206]. Noncommutative L^p-spaces generalize Schatten classes.

Let M be a semifinite von Neumann algebra of bounded linear operators defined on $\mathcal{H}$ and let τ be a semifinite normal faithful trace on M. Note that $(\mathcal{B}(\mathcal{H}), \mathrm{Tr})$ is one of examples of (M, τ). We denote the subset of self-adjoint elements of M by M_{sa}.

We use the notation $X\eta M$ for a closed densely defined operator X affiliated with M, and $H\eta M_{sa}$ for a self-adjoint operator $H\eta M$ (i.e., all the spectral projections of H are elements of M). Let $S(M, \tau)$ denote the set of τ-measurable operators affiliated with M. We recall that $H\eta M_{sa}$ is τ-measurable if given $\epsilon > 0$ there exists a projection $P \in M$ such that $P(\mathcal{H}) \subset \mathrm{dom}(H)$ and $\tau(I - P) < \epsilon$ [79, Definition 1.2].

Let $\mu_t(X)$ denote the tth generalized s-number [79, Definition 2.1] of $X \in S(M, \tau)$. By [79, Proposition 2.2],

$$\mu_t(X) = \inf\{s \geqslant 0 : \ \tau(E_{|X|}(s, \infty)) \leqslant t\}.$$

Further properties of generalized s-numbers can be found in [79]. For instance, if $(M, \tau) = (\mathcal{B}(\mathcal{H}), \mathrm{Tr})$, then $S(\mathcal{B}(\mathcal{H}), \mathrm{Tr}) = \mathcal{B}(\mathcal{H})$ and the generalized s-numbers coincide with the singular values of the operators, namely,

$$\mu_t(X) = s_n(X), \quad t \in [n-1, n), \ n \in \mathbb{N}.$$

An operator $X \in M$ is said to be τ-compact if and only if

$$\lim_{t \to \infty} \mu_t(X) = 0.$$

An operator $X \in M$ is said to be τ-(Breuer-)Fredholm if the projections on $\ker X$ and $\ker X^*$ are τ-finite and there exists a τ-finite projection $P \in M$ such that $\mathrm{ran}(I - P) \subseteq \mathrm{ran}(X)$. If an operator has τ-compact resolvent, then it is τ-Fredholm.

The noncommutative L^p-space, $1 \leqslant p < \infty$, associated with (M, τ) is

$$L^p(M, \tau) := \{X\eta M : \ \|X\|_p := \tau(|X|^p)^{1/p} < \infty\}.$$

This space can also be described as

$$L^p(M, \tau) = \{X \in S(M, \tau) : \ \mu(X) \in L^p(0, \infty)\},$$

where $(L^p(0, \infty), \|\cdot\|_p)$ is the usual Lebesgue space, and $\mu(X)$ denotes the function $t \mapsto \mu_t(X)$. In this case we have

$$\|X\|_p = \|\mu(X)\|_p, \quad X \in L^p(\mathcal{M}, \tau).$$

An example of a noncommutative L^p-space is the Schatten ideal $\mathcal{S}^p = L^p(\mathcal{B}(\mathcal{H}), \text{Tr})$. We denote

$$L^\infty(\mathcal{M}, \tau) := \mathcal{M}$$

and let $\|\cdot\|_\infty = \|\cdot\|$ stand for the operator norm. The Hölder inequality (2.4.1) extends to the setting of noncommutative L^p-spaces:

$$\|XY\|_r \leqslant \|X\|_p \|Y\|_q,$$

where $1 \leqslant p, q, r \leqslant \infty$ are such that $\frac{1}{r} = \frac{1}{p} + \frac{1}{q}$ and $X \in L^p(\mathcal{M}, \tau)$, $Y \in L^q(\mathcal{M}, \tau)$. We have the duality

$$\left(L^p(\mathcal{M}, \tau)\right)^* = L^q(\mathcal{M}, \tau),$$

where $1 \leqslant p < \infty$, $\frac{1}{p} + \frac{1}{q} = 1$.

The weak noncommutative L^p-space, $1 \leqslant p < \infty$, associated with $(\mathcal{M}, \tau)$ is the space

$$L^{p,\infty}(\mathcal{M}, \tau) := \left\{ X \in S(\mathcal{M}, \tau) : \|X\|'_{L^{p,\infty}} := \sup_{t \geqslant 0} t^{\frac{1}{p}} \mu_t(X) < +\infty \right\}.$$

If $1 \leqslant p < \infty$, then the space $L^{p,\infty}(\mathcal{M}, \tau)$ equipped with the quasi-norm $\|\cdot\|'_{L^{p,\infty}}$ given above becomes a quasi-Banach space (see, e.g., [123, Example 2.6.10] and [68]). For $1 < p < \infty$, there exists a norm $\|\cdot\|_{L^{p,\infty}}$ on $L^{p,\infty}(\mathcal{M}, \tau)$ given by

$$\|X\|_{L^{p,\infty}} := \sup_{t > 0} t^{\frac{1}{p}-1} \int_0^t \mu_s(X) ds, \quad X \in L^{p,\infty}(\mathcal{M}, \tau),$$

which satisfies

$$\|\cdot\|'_{L^{p,\infty}} \leqslant \|\cdot\|_{L^{p,\infty}} \leqslant \frac{p}{p-1} \|\cdot\|'_{L^{p,\infty}}. \tag{2.6.1}$$

In the special case when $\mathcal{M}$ is the commutative von Neumann algebra $L^\infty(0, \infty)$ equipped with the normal semifinite trace given by integration with respect to the Lebesgue measure, the space $L^{p,\infty}(\mathcal{M}, \tau)$ coincides with the classical commutative weak L^p-space $L^{p,\infty}(0, \infty)$ (see, e.g., [68, 117, 122]).

The following result is the Hölder type inequality for the quasi-norm $\|\cdot\|'_{L^{p,\infty}}$ and the norm $\|\cdot\|_{L^{p,\infty}}$.

Lemma 2.6.1 *Let $m \in \mathbb{N}$ and let $1 \leqslant p, p_1, \ldots, p_m < \infty$ be such that $\frac{1}{p_1} + \cdots + \frac{1}{p_m} = \frac{1}{p}$.*

(i) For all $X_j \in L^{p_j, \infty}(\mathcal{M}, \tau)$, $1 \leqslant j \leqslant m$,

$$\|X_1 \cdot \ldots \cdot X_m\|'_{L^{p,\infty}} \leqslant m^{\frac{1}{p}} \prod_{j=1}^{m} \|X_j\|_{L^{p_j,\infty}}.$$

(ii) If $1 < p, p_1, \ldots p_m < \infty$, then for all $X_j \in L^{p_j, \infty}(\mathcal{M}, \tau)$, $1 \leqslant j \leqslant m$,

$$\|X_1 \cdot \ldots \cdot X_m\|_{L^{p,\infty}} \leqslant \frac{p}{p-1} m^{\frac{1}{p}} \prod_{j=1}^{m} \|X_j\|_{L^{p_j,\infty}}.$$

Proof

(i) It follows from [123, Corollary 2.3.16] that for all $t > 0$,

$$\mu_t(X_1 \cdot \ldots \cdot X_m) \leqslant \prod_{j=1}^{m} \mu_{\frac{t}{m}}(X_j) \leqslant \prod_{j=1}^{m} \left(\frac{t}{m}\right)^{-\frac{1}{p_j}} \|X_j\|'_{L^{p_j,\infty}}$$

$$= \left(\frac{t}{m}\right)^{-\frac{1}{p}} \prod_{j=1}^{m} \|X_j\|'_{L^{p_j,\infty}}.$$

Hence,

$$t^{\frac{1}{p}} \mu_t(X_1 \cdot \ldots \cdot X_m) \leqslant m^{\frac{1}{p}} \prod_{j=1}^{m} \|X_j\|'_{L^{p_j,\infty}}.$$

Taking the supremum over $t > 0$ and using (2.6.1) proves *(i)*.

(ii) It follows from (2.6.1) and (i) that

$$\|X_1 \cdot \ldots \cdot X_m\|_{L^{p,\infty}} \leqslant \frac{p}{p-1} \|X_1 \cdot \ldots \cdot X_m\|'_{L^{p,\infty}} \leqslant \frac{p}{p-1} m^{\frac{1}{p}} \prod_{j=1}^{m} \|X_j\|_{L^{p_j,\infty}},$$

completing the proof. $\square$

Recall that for $1 \leqslant r, q \leqslant \infty$, the space $(L^q + L^r)(\mathcal{M}, \tau)$ is defined as

$$(L^q + L^r)(\mathcal{M}, \tau) := \{X \in S(\mathcal{M}, \tau) : \|X\|_{(L^q+L^r)(\mathcal{M},\tau)} < \infty\},$$

where

$$\|X\|_{(L^q+L^r)(\mathcal{M},\tau)} := \inf\{\|Y\|_q + \|Z\|_r : \ X = Y+Z, \ Y \in L^q(\mathcal{M}, \tau), \ Z \in L^r(\mathcal{M}, \tau)\}.$$

We shall sometimes suppress the algebra $(\mathcal{M}, \tau)$ from the notations. For example, we shall write $\| \cdot \|_{L^q + L^r}$ rather than $\| \cdot \|_{(L^q + L^r)(\mathcal{M}, \tau)}$. This should not cause any confusion.

Observe that if $1 < r < p < q < \infty$, then

$$L^p(\mathcal{M}, \tau) \subset L^{p, \infty}(\mathcal{M}, \tau) \subset (L^q + L^r)(\mathcal{M}, \tau)$$

(see, e.g., [122, Proposition 2.b.9 and p. 143]).

The following simple lemma is essential for a subsequent exposition.

Lemma 2.6.2 *Let $k \in \mathbb{N}$ and $k < r < q < \infty$. Let T be a multilinear operator with domain* $\mathrm{dom}(T) \subset S(\mathcal{M}, \tau)^{\times k}$. *Suppose that for all $1 < p, p_j < \infty$, $j = 1, \ldots, k$, satisfying $\frac{1}{p} = \frac{1}{p_1} + \cdots + \frac{1}{p_k}$ there exists a constant $c_p > 0$ such that*

$$\| T : L^{p_1}(\mathcal{M}, \tau) \times \ldots \times L^{p_k}(\mathcal{M}, \tau) \to L^p(\mathcal{M}, \tau) \| \leqslant c_p. \tag{2.6.2}$$

Then,

$$T \in \mathcal{B}_k((L^q + L^r)(\mathcal{M}, \tau)^{\times k}, (L^{\frac{q}{k}} + L^{\frac{r}{k}})(\mathcal{M}, \tau)).$$

Proof By the definition of the space $L^q + L^r$, for $X_1, \ldots, X_k \in (L^q + L^r)(\mathcal{M}, \tau)$, there are $Y_j \in L^q(\mathcal{M}, \tau)$, $Z_j \in L^r(\mathcal{M}, \tau)$ such that $X_j = Y_j + Z_j$, $j = 1, \ldots, k$. Hence,

$$\| T(X_1, \ldots, X_k) \|_{L^{\frac{q}{k}} + L^{\frac{r}{k}}} = \| T(Y_1 + Z_1, \ldots, Y_k + Z_k) \|_{L^{\frac{q}{k}} + L^{\frac{r}{k}}}$$

$$= \left\| \sum_{\mathscr{A} \subset \{1, \ldots, k\}} T(X_{1, \mathscr{A}}, \ldots, X_{k, \mathscr{A}}) \right\|_{L^{\frac{q}{k}} + L^{\frac{r}{k}}}$$

$$\leqslant \sum_{\mathscr{A} \subset \{1, \ldots, k\}} \| T(X_{1, \mathscr{A}}, \ldots, X_{k, \mathscr{A}}) \|_{L^{\frac{q}{k}} + L^{\frac{r}{k}}}, \tag{2.6.3}$$

where

$$X_{j, \mathscr{A}} = \begin{cases} Y_j, & j \in \mathscr{A} \\ Z_j, & j \notin \mathscr{A} \end{cases}, \quad j = 1, \ldots, k.$$

Fix $\mathscr{A} \subset \{1, \ldots, k\}$. If $1 < p_{\mathscr{A}} < \infty$ is such that

$$\frac{1}{p_{\mathscr{A}}} = \frac{|\mathscr{A}|}{r} + \frac{k - |\mathscr{A}|}{q},$$

then

$$\frac{r}{k} \leqslant p_{\mathscr{A}} \leqslant \frac{q}{k}$$

and, therefore, $L^{p_{\mathscr{A}}}(M, \tau) \subset (L^{\frac{q}{k}} + L^{\frac{r}{k}})(M, \tau)$. Thus,

$$\|T(X_{1,\mathscr{A}}, \ldots, X_{k,\mathscr{A}})\|_{L^{\frac{q}{k}} + L^{\frac{r}{k}}} \leqslant \text{const } \|T(X_{1,\mathscr{A}}, \ldots, X_{k,\mathscr{A}})\|_{L^{p_{\mathscr{A}}}}. \qquad (2.6.4)$$

Using (2.6.2), we obtain

$$\|T(X_{1,\mathscr{A}}, \ldots, X_{k,\mathscr{A}})\|_{L^{p_{\mathscr{A}}}} \leqslant c_{p_{\mathscr{A}}} \prod_{j \in \mathscr{A}} \|Y_j\|_{L^q} \prod_{j \notin \mathscr{A}} \|Z_j\|_{L^r}$$

$$\leqslant c_{p_{\mathscr{A}}} \prod_{j=1}^{k} (\|Y_j\|_{L^q} + \|Z_j\|_{L^r}). \qquad (2.6.5)$$

Combining (2.6.3)–(2.6.5) implies

$$\|T(X_1, \ldots, X_k)\|_{L^{\frac{q}{k}} + L^{\frac{r}{k}}} \leqslant \text{const} \sum_{\mathscr{A} \subset \{1, \ldots, k\}} \|T(X_{1,\mathscr{A}}, \ldots, X_{k,\mathscr{A}})\|_{L^{p_{\mathscr{A}}}}$$

$$\leqslant \text{const} \sum_{\mathscr{A} \subset \{1, \ldots, k\}} c_{p_{\mathscr{A}}} \prod_{j=1}^{k} (\|Y_j\|_{L^q} + \|Z_j\|_{L^r}).$$

Taking the infimum over all the representations $X_j = Y_j + Z_j$, $Y_j \in L^q(M, \tau)$, $Z_j \in L^r(M, \tau)$, $j = 1, \ldots, k$, completes the proof. $\qquad\square$

2.7   The Haagerup L^p-Space

In this section we recall the construction of noncommutative L^p-spaces associated with an arbitrary von Neumann algebra. We use Haagerup's definition [87], and Terp's exposition of the subject [206]. The basics on von Neumann algebras and Tomita's modular theory can be found in [96].

Let M be an arbitrary von Neumann algebra with a faithful normal semifinite weight ϕ_0. We consider the one-parameter modular automorphism group $\sigma^{\phi_0} = \{\sigma_t^{\phi_0}\}_{t \in \mathbb{R}}$ (associated with ϕ_0) on M and obtain a semifinite crossed product von Neumann algebra

$$\mathcal{N} := M \rtimes_{\sigma^{\phi_0}} \mathbb{R}, \qquad (2.7.1)$$

which admits the canonical semifinite trace τ and a trace-scaling dual action $\theta = \{\theta_s\}_{s \in \mathbb{R}}$ such that

$$\tau \circ \theta_s = e^{-s}\tau \quad \text{for all } s \in \mathbb{R}.$$

The original von Neumann algebra $\mathcal{M}$ can be identified with a θ-invariant von Neumann subalgebra $L^\infty_{Haag}(\mathcal{M})$ of $\mathcal{N}$. For $1 \leqslant p < \infty$, the noncommutative Haagerup L^p-space $L^p_{Haag}(\mathcal{M})$ is defined by

$$L^p_{Haag}(\mathcal{M}) := \{X \in S(\mathcal{N}, \tau) : \theta_s(X) = e^{-\frac{s}{p}} X \text{ for all } s \in \mathbb{R}\}.$$

It is known from [206, Part II, Theorem 7] that there is a linear bijection $\psi \mapsto X_\psi$ between the predual space $\mathcal{M}_*$ and $L^1_{Haag}(\mathcal{M})$. Due to this correspondence we define the trace $\mathrm{tr} : L^1_{Haag}(\mathcal{M}) \to \mathbb{C}$ by

$$\mathrm{tr}(X_\psi) := \psi(I), \quad X_\psi \in L^1_{Haag}(\mathcal{M}). \tag{2.7.2}$$

Given any $X \in L^p_{Haag}(\mathcal{M})$, $1 \leqslant p < \infty$, we have the polar decomposition $X = U|X|$, where $|X|$ is a positive operator in $L^p_{Haag}(\mathcal{M})$ and U is a partial isometry contained in $\mathcal{M}$. It is established in [206, Proposition 12] that $|X|^p \in L^1_{Haag}(\mathcal{M})$. Thus, we can define a Banach norm (see [206, Corollary 27]) on $L^p_{Haag}(\mathcal{M})$ by setting

$$\|X\|_{L^p_{Haag}} := \mathrm{tr}(|X|^p)^{\frac{1}{p}}, \quad X \in L^p_{Haag}(\mathcal{M}). \tag{2.7.3}$$

The following lemma due to [79, Lemma 1.7] shows that if $X \in L^p_{Haag}(\mathcal{M})$, then

$$\|X\|_{L^p_{Haag}} = \|X\|'_{L^{p,\infty}} \tag{2.7.4}$$

and, therefore, $L^p_{Haag}(\mathcal{M})$ is a closed linear subspace in $L^{p,\infty}(\mathcal{N}, \tau)$.

Lemma 2.7.1 *Let* $1 \leqslant p < \infty$. *If* $X \in L^p_{Haag}(\mathcal{M})$, *then*

$$\mu_t(X) = \|X\|_{L^p_{Haag}} \cdot t^{-\frac{1}{p}}, \quad t > 0.$$

The following version of the Hölder inequality is proved, for instance, in [206, Theorem 23].

Lemma 2.7.2 *Let* $k \in \mathbb{N}$ *and* $1 \leqslant p, p_j \leqslant \infty$, $j = 1, \ldots, k$, *be such that* $\frac{1}{p_1} + \cdots + \frac{1}{p_n} = \frac{1}{p}$. *For every* $X_j \in L^{p_j}_{Haag}(\mathcal{M})$, $1 \leqslant j \leqslant k$,

$$\|X_1 \cdot \ldots \cdot X_k\|_{L^p_{Haag}} \leqslant \|X_1\|_{L^{p_1}_{Haag}} \cdot \ldots \cdot \|X_k\|_{L^{p_k}_{Haag}}. \tag{2.7.5}$$

Let now $\mathcal{M}$ be a semifinite von Neumann algebra and τ_0 a faithful normal semifinite trace on $\mathcal{M}$. Let $\mathcal{M} \bar{\otimes} L^\infty(\mathbb{R})$ be the von Neumann algebra tensor product of M and $L^\infty(\mathbb{R})$ acting on the Hilbert space tensor product $\mathcal{H}\bar\otimes L^2(\mathbb{R})$ equipped with the tensor product trace $\tau' := \tau_0 \otimes \nu$, where ν is the trace on $L^\infty(\mathbb{R})$, given by

$$\nu(f) = \int_{\mathbb{R}} f(s)e^{-s}ds, \quad 0 \leqslant f \in L^\infty(\mathbb{R}).$$

Recall that τ' is the unique faithful normal semifinite trace on $M\bar\otimes L^\infty(\mathbb{R})$ satisfying $\tau'(X \otimes f) = \tau_0(X)\nu(f)$, $X \in M$, $f \in L^\infty(\mathbb{R})$.

It is known that in the case when M is a semifinite von Neumann algebra there exists a trace preserving $*$-isomorphism between the crossed product von Neumann algebra $(\mathcal{N}, \tau)$ defined in (2.7.1) and $(M\bar\otimes L^\infty(\mathbb{R}), \tau')$ (see [209, Part II, Proposition 4.2]). We identify $(\mathcal{N}, \tau)$ with $(M\bar\otimes L^\infty(\mathbb{R}), \tau')$. It is also known that for all $X \in M$, $f \in L^\infty(\mathbb{R})$,

$$\theta_s(X \otimes f) = X \otimes l_s(f), \quad s \in \mathbb{R},$$

where l_s is the left translation by s (see, e.g., [209, Part II, Proposition 4.2]).

The following result is well-known (see, e.g., [87, Theorem 2.1], [206, p. 62]).

Theorem 2.7.3 *Let $1 \leqslant p \leqslant \infty$ and $\zeta_p(t) = e^{\frac{t}{p}}$, $t \in \mathbb{R}$. Let M be a semifinite von Neumann algebra equipped with a faithful normal semifinite trace τ_0. The following assertions hold.*

(i) *The operator $X \otimes \zeta_p$ affiliated with $M\bar\otimes L^\infty(\mathbb{R})$ is τ'-measurable for all $X \in L^p(M, \tau_0)$.*

(ii) *$X \otimes \zeta_p \in L^p_{Haag}(M)$, for all $X \in L^p(M, \tau_0)$.*

(iii) *The mapping*

$$X \mapsto X \otimes \zeta_p, \quad X \in L^p(M, \tau_0)$$

is an isometry from $L^p(M, \tau_0)$ into $L^p_{Haag}(M)$.

2.8 Symmetrically Normed Ideals

The definitions of objects recalled in this section can be found in, for instance, [123]. For a more detailed treatment of the subject we refer the reader to [57, 85, 183].

Let M be a semifinite von Neumann algebra. Let $\mathcal{I}$ be a symmetrically normed ideal of M with norm $\|\cdot\|_\mathcal{I}$, that is, $\mathcal{I}$ is a two sided Banach ideal with respect to the norm $\|\cdot\|_\mathcal{I}$, which satisfies the properties

(i) $A \in M$, $B \in \mathcal{I}$, $0 \leqslant A \leqslant B$ implies $A \in \mathcal{I}$ and $\|A\|_\mathcal{I} \leqslant \|B\|_\mathcal{I}$,

(ii) there is a constant $c_1 > 0$ such that $\|B\| \leqslant c_1 \|B\|_\mathcal{I}$ for every $B \in \mathcal{I}$,

(iii) for all $A, C \in M$ and $B \in \mathcal{I}$, we have $\|ABC\|_\mathcal{I} \leqslant \|A\| \|B\|_\mathcal{I} \|C\|$.

A simple example of $\mathcal{I}$ is the ideal

$$\mathcal{L}^p(\mathcal{M}, \tau) := L^p(\mathcal{M}, \tau) \cap \mathcal{M}$$

equipped with the norm $\|\cdot\|_{\mathcal{I}} = \max\{\|\cdot\|, \|\cdot\|_p\}$. Another example is the dual Macaev (also called classical Dixmier-Macaev) ideal

$$\mathcal{L}^{(1,\infty)} := \left\{ A \in \mathcal{S}^\infty : \|A\|_{(1,\infty)} := \sup_{n \in \mathbb{N}} \frac{1}{\log(1+n)} \sum_{k=1}^{n} s_k(A) < \infty \right\}.$$

To get another class of examples we let ψ be a concave function satisfying

$$\lim_{t \to 0^+} \psi(t) = 0, \qquad \lim_{t \to \infty} \psi(t) = \infty.$$

Then the Marcinkiewicz (also called Lorentz) ideal $\mathcal{I}_\psi$ associated with a σ-finite, semifinite von Neumann algebra factor $\mathcal{M}$ and its ideal norm $\|\cdot\|_{\mathcal{I}_\psi}$ are defined by

$$\mathcal{I}_\psi = \left\{ A \in \mathcal{M} : \|A\|_{\mathcal{I}_\psi} := \max \left\{ \|A\|, \sup_{t>0} \frac{1}{\psi(t)} \int_0^t \mu_s(A)\, ds < \infty \right\} \right\}.$$

Let $\tau_{\mathcal{I}}$ be a trace on a symmetrically normed ideal $\mathcal{I}$, that is, $\tau_{\mathcal{I}}$ is a linear functional on $\mathcal{I}$ satisfying the unitary invariance

$$\tau_{\mathcal{I}}(UAU^*) = \tau_{\mathcal{I}}(A) \text{ for all } A \in \mathcal{I}, \text{ unitary } U \in \mathcal{M}.$$

The unitary invariance of the trace $\tau_{\mathcal{I}}$ implies its cyclicity property

$$\tau_{\mathcal{I}}(AB) = \tau_{\mathcal{I}}(BA) \text{ for all } A \in \mathcal{M}, B \in \mathcal{I}$$

(see, e.g., [96, Proposition 8.1.1]). We also assume that $\tau_{\mathcal{I}}$ is positive, that is,

$$\tau_{\mathcal{I}}(A) \geqslant 0 \text{ for all } A \in \mathcal{I}, A \geqslant 0$$

and that $\tau_{\mathcal{I}}$ is bounded (equivalently, continuous) with respect to the ideal norm $\|\cdot\|_{\mathcal{I}}$, that is, there is a constant $c_2 > 0$ such that

$$|\tau_{\mathcal{I}}(A)| \leqslant c_2 \|A\|_{\mathcal{I}} \text{ for all } A \in \mathcal{I}.$$

Examples of $(\mathcal{I}, \tau_{\mathcal{I}})$ include $(\mathcal{S}^1, \mathrm{Tr})$, $(\mathcal{L}^1(\mathcal{M}, \tau), \tau)$, $(\mathcal{L}^{(1,\infty)}, \mathrm{Tr}_{\omega_d})$, and $(\mathcal{I}_\psi, \tau_{\omega_c})$, where Tr_{ω_d} is the Dixmier trace corresponding to a generalized limit ω_d on ℓ^∞ and τ_{ω_c} is the Dixmier trace corresponding to a dilation invariant Banach limit ω_c on $L^\infty(0, \infty)$.

Given a symmetrically normed ideal I equipped with a continuous trace τ_I, we consider the root ideals

$$I^{1/p} = \{A \in M : |A|^p \in I\},$$

$p \in \mathbb{N}$, whose elements satisfy the Hölder inequality

$$\|AB\|_I \leqslant \|A\|_{I^{1/p}} \|B\|_{I^{1/q}}, \quad \frac{1}{p} + \frac{1}{q} = 1,$$

where

$$\|A\|_{I^{1/p}} := (\| |A|^p \|_I)^{1/p}$$

(see [77, 200]).

2.9   Traces on $L^{1,\infty}(M, \tau)$

Let M be a semifinite von Neumann algebra equipped with a normal faithful semifinite trace τ. In this section we adopt some terminology from the theory of singular traces on symmetric operator spaces (see [123]) to traces on the noncommutative weak L^1-space $L^{1,\infty}(M, \tau)$ (see also [73, Section 6]).

A trace ϕ on $L^{1,\infty}(M, \tau)$ is a linear functional, which is unitarily invariant, that is, $\phi : L^{1,\infty}(M, \tau) \to \mathbb{C}$ satisfies $\phi(UXU^*) = \phi(X)$ for all $X \in L^{1,\infty}(M, \tau)$ and all unitaries $U \in M$. A trace ϕ on $L^{1,\infty}(M, \tau)$ is said to be normalized if $\phi(X) = 1$ for every $0 \leqslant X \in L^{1,\infty}(M, \tau)$ with $\mu_t(X) = t^{-1}$, $t > 0$.

Fix a free ultrafilter ω on $\mathbb{N}$. The limit with respect to the ultrafilter ω is denoted by $\lim_{n \to \omega}$. In the following lemma we introduce a particular trace on $L^{1,\infty}(M, \tau)$, which is essential in the proof of the main result in Sect. 5.8.

Lemma 2.9.1 *The functional* $\phi : (L^{1,\infty}(M, \tau))_+ \to \mathbb{R}_+$, *given by*

$$\phi(X) := \lim_{n \to \omega} \frac{1}{\log(1+n)} \int_1^n \mu_t(X)\, dt, \quad 0 \leqslant X \in L^{1,\infty}(M, \tau), \qquad (2.9.1)$$

extends to a positive normalized trace on $L^{1,\infty}(M, \tau)$.

Proof First we show that ϕ is a linear functional on all positive elements of $L^{1,\infty}(M, \tau)$. The equality

$$\phi(\alpha X) = \alpha\phi(X), \quad \alpha > 0, \quad 0 \leqslant X \in L^{1,\infty}(M, \tau),$$

is obvious. Next we show that $\phi(X + Y) = \phi(X) + \phi(Y)$ for $0 \leqslant X, Y \in L^{1,\infty}(M, \tau)$.

Let $n > 1$ be fixed. By [97, Lemma 8.4], there exists $\lambda \in \mathbb{N}$ satisfying

$$\int_{\lambda a}^{n} \mu_t(X + Y) dt \leqslant \int_{a}^{n} (\mu_t(X) + \mu_t(Y)) dt,$$

for all $\lambda a < n$. Taking $a = 1/\lambda$ we infer that

$$\int_{1}^{n} \mu_t(X + Y) dt \leqslant \int_{1/\lambda}^{n} (\mu_t(X) + \mu_t(Y)) dt$$

$$= \int_{1}^{n} (\mu_t(X) + \mu_t(Y)) dt + \int_{1/\lambda}^{1} (\mu_t(X) + \mu_t(Y)) dt.$$

Dividing both parts of the latter inequality by $\log(1 + n)$ and taking $\lim_{n \to \omega}$ imply

$$\phi(X + Y) \leqslant \phi(X) + \phi(Y).$$

Conversely, by [123, Lemma 3.4.4], there exists $\lambda \in \mathbb{N}$ such that

$$\int_{\lambda a}^{n} (\mu_t(X) + \mu_t(Y)) dt \leqslant 2 \int_{a}^{n} \mu_{2t}(X + Y) dt, \quad \lambda a < n.$$

Without loss of generality, we may assume that $\lambda > 2$. Setting $a = 1/\lambda$ implies

$$\int_{1}^{n} (\mu_t(X) + \mu_t(Y)) dt \leqslant 2 \int_{1/\lambda}^{n} \mu_{2t}(X+Y) dt = \int_{2/\lambda}^{2n} \mu_t(X+Y) dt \qquad (2.9.2)$$

$$= \int_{2/\lambda}^{1} \mu_t(X+Y) dt + \int_{1}^{n} \mu_t(X+Y) dt + \int_{n}^{2n} \mu_t(X+Y) dt.$$

Observe that

$$\int_{n}^{2n} \mu_t(X + Y) dt \leqslant \|X + Y\|'_{L^{1,\infty}} \int_{n}^{2n} t^{-1} dt = \|X + Y\|'_{L^{1,\infty}} \log 2.$$

Dividing (2.9.2) by $\log(1 + n)$ and taking $\lim_{n \to \omega}$ imply

$$\phi(X) + \phi(Y) \leqslant \phi(X + Y).$$

Thus, ϕ is additive and positively homogeneous on

$$(L^{1,\infty}(\mathcal{M}, \tau))_+ := \{X \in L^{1,\infty}(\mathcal{M}, \tau) : X \geqslant 0\}.$$

It extends to $L^{1,\infty}(\mathcal{M}, \tau)$ by linearity, and we denote this extension by ϕ. Since $\mu(UXU^*) = \mu(X)$ for all $X \in L^{1,\infty}(\mathcal{M}, \tau)$ and all unitaries $U \in \mathcal{M}$, it follows that

ϕ is unitarily invariant. The fact that ϕ is positive and normalized follows directly from definition (2.9.1). $\square$

Lemma 2.9.2 *The extension of ϕ defined in (2.9.1) is a bounded linear functional on $L^{1,\infty}(\mathcal{M}, \tau)$.*

Proof If $X \in (L^{1,\infty}(\mathcal{M}, \tau))_+$, then

$$|\phi(X)| \leqslant \|X\|'_{L^{1,\infty}} \lim_{n \to \omega} \frac{1}{\log(1+n)} \int_1^n t^{-1} dt = \|X\|'_{L^{1,\infty}}.$$

Let now $X \in L^{1,\infty}(\mathcal{M}, \tau)$. Then,

$$X = (\operatorname{Re} X)_+ - (\operatorname{Re} X)_- + i(\operatorname{Im} X)_+ - i(\operatorname{Im} X)_-.$$

By [98, Chapter 1],

$$\|(\operatorname{Re} X)_\pm\|'_{L^{1,\infty}}, \quad \|(\operatorname{Im} X)_\pm\|'_{L^{1,\infty}} \leqslant c_0 \|X\|'_{L^{1,\infty}},$$

where c_0 is the modulus of concavity of the quasi-norm $\| \cdot \|'_{L^{1,\infty}}$. Therefore,

$$|\phi(X)| \leqslant \left(\max\{\phi((\operatorname{Re} X)_+)^2, \phi((\operatorname{Re} X)_-)^2\} + \max\{\phi((\operatorname{Im} X)_+)^2, \phi((\operatorname{Im} X)_-)^2\} \right)^{\frac{1}{2}}$$
$$\leqslant \sqrt{2}\, c_0 \|X\|'_{L^{1,\infty}}.$$

Lemma 2.9.3 *If $X \in L^{1,\infty}(\mathcal{M}, \tau)$ is such that $\tau(E_{|X|}(0, \infty)) < \infty$, then $\phi(X) = 0$, where ϕ is defined in (2.9.1).*

Proof Let $c := \tau(E_{|X|}(0, \infty))$ for some $c > 0$. Since $\mu(X)$ is the right inverse function of $t \to \tau(E_{|X|}(t, \infty))$, it follows that $\mu_c(X) = 0$. Therefore,

$$\int_1^n \mu_t(X)\, dt = \int_1^c \mu_t(X)\, dt, \quad \text{for all } n \geqslant c,$$

implying $\phi(X) = 0$. $\square$

The proof of the following lemma immediately follows from Lemma 2.7.1 and existence of the Jordan decomposition of elements from $L^1_{Haag}(\mathcal{M})$ (see [15, Theorem 6]).

Lemma 2.9.4 *Let $(\mathcal{N}, \tau)$ be the crossed product von Neumann algebra defined in (2.7.1) and let tr be the trace on $L^1_{Haag}(\mathcal{M})$ defined by (2.7.2). Then, for any normalized trace ϕ on $L^{1,\infty}(\mathcal{N}, \tau)$,*

$$\phi(X) = \operatorname{tr}(X), \quad X \in L^1_{Haag}(\mathcal{M}).$$

2.10 Banach Spaces and Spectral Operators

In this section we collect facts on operators on Banach spaces for which double operator integrals can be constructed and admit results similar to those on a Hilbert space.

Let μ be a complex Borel measure on a measurable space (Ω, Σ) and X a Banach space. A function $f : \Omega \to X$ is called μ-measurable if there exists a sequence of X-valued simple functions converging to f μ-almost everywhere. For Banach spaces X and Y, we say that a function $f : \Omega \to \mathcal{B}(X, Y)$ is strongly measurable if $\omega \mapsto f(\omega)x$ is a μ-measurable mapping $\Omega \to Y$ for each $x \in X$.

A Banach space X is said to have a bounded approximation property if there exists $M \geqslant 1$ such that for every compact subset $K \subseteq X$ and $\epsilon > 0$, there exists $X \in X^* \otimes X$ with $\|X\|_{X \to X} \leqslant M$ and $\sup_{x \in K} \|Xx - x\|_X < \epsilon$. If X and Y are Banach spaces such that X is separable and either X or Y has a bounded approximation property, then every $T \in \mathcal{B}(X, Y)$ is a limit in the strong operator topology of a norm bounded sequence of finite rank operators [173, Lemma 3.2].

Let $(Z, \| \cdot \|_Z)$ be a Banach space which is continuously embedded in $\mathcal{B}(X, Y)$. Following [210], we say that Z has the strong convex compactness property if for every finite measure space (Ω, Σ, μ) and every strongly measurable bounded $f : \Omega \to Z$, the operator $T \in \mathcal{B}(X, Y)$ defined by

$$Tx := \int_\Omega f(\omega)x \, d\mu(\omega), \qquad x \in X,$$

belongs to Z with

$$\|T\|_Z \leqslant \inf \int_\Omega g(\omega) \, d\mu(\omega),$$

where the infimum is taken over all measurable functions $g : \Omega \to [0, \infty]$ such that $\|f(\omega)\|_Z \leqslant g(\omega)$ for $\omega \in \Omega$. Any separable Z has the strong convex compactness property. The subspaces of compact and weakly compact operators in $\mathcal{B}(X, Y)$ have the strong convex compactness property, but not all subspaces of $\mathcal{B}(X, Y)$ do. If N is a semifinite von Neumann algebra on a separable Hilbert space H with a faithful normal semifinite trace τ and F is a rearrangement invariant Banach function space with the Fatou property, then $N \cap F(N, \tau)$ has the strong convex compactness property (see [173] for more details).

Let X and Y be Banach spaces and I a Banach space which is continuously embedded in $\mathcal{B}(X, Y)$. We say that $(I, \| \cdot \|_I)$ is a Banach ideal in $\mathcal{B}(X, Y)$ if

- $R \in \mathcal{B}(Y)$, $X \in I$, and $T \in \mathcal{B}(X)$ imply $RXT \in I$ and

$$\|RXT\|_I \leqslant \|R\|_{Y \to Y} \|X\|_I \|T\|_{X \to X};$$

- $X^* \otimes Y \subseteq I$ and $\|x^* \otimes y\|_I = \|x^*\|_{X^*} \|y\|_Y$ for all $x^* \in X^*$ and $y \in Y$.

For separable $\mathcal{X}$ and $\mathcal{Y}$, any maximal Banach ideal (for the definition see, e.g., [152]) in $\mathcal{B}(\mathcal{X}, \mathcal{Y})$ has the strong convex compactness property. This includes a large class of operator ideals, such as the ideal of absolutely p-summing operators, the ideal of integral operators, etc.

Below we summarize some of the basics of scalar type operators, as taken from [72].

Let $\mathcal{X}$ be a Banach space and $\mathfrak{B}$ a Borel σ-algebra. A spectral measure on $\mathcal{X}$ is a map $E : \mathfrak{B} \to \mathcal{B}(\mathcal{X})$ such that the following hold:

- $E(\varnothing) = 0$ and $E(\mathbb{C}) = I_{\mathcal{X}}$;
- $E(\varsigma_1 \cap \varsigma_2) = E(\varsigma_1) E(\varsigma_2)$ for all $\varsigma_1, \varsigma_2 \in \mathfrak{B}$;
- $E(\varsigma_1 \cup \varsigma_2) = E(\varsigma_1) + E(\varsigma_2) - E(\varsigma_1) E(\varsigma_2)$ for all $\varsigma_1, \varsigma_2 \in \mathfrak{B}$;
- E is σ-additive in the strong operator topology.

Note that these conditions imply that E is projection-valued. Moreover, by [72, Corollary XV.2.4] there exists a constant K such that

$$\|E(\varsigma)\|_{\mathcal{B}(\mathcal{X})} \leqslant K, \qquad \varsigma \in \mathfrak{B}. \tag{2.10.1}$$

An operator $A \in \mathcal{B}(\mathcal{X})$ is a spectral operator if there exists a spectral measure E on $\mathcal{X}$ such that $AE(\varsigma) = E(\varsigma)A$ and $\sigma(A, E(\varsigma)\mathcal{X}) \subseteq \bar{\varsigma}$ for all $\varsigma \in \mathfrak{B}$, where $\sigma(A, E(\varsigma)\mathcal{X})$ denotes the spectrum of A in the space $E(\varsigma)\mathcal{X}$. For a spectral operator A, we let $\mathrm{spec}(A)$ denote the minimal constant K occurring in (2.10.1) and call $\mathrm{spec}(A)$ the spectral constant of A, which is well-defined since the spectral measure E associated with A is unique (see [72, Corollary XV.3.8]). By [72, Corollary XV.3.5], $E(\sigma(A)) = I_{\mathcal{X}}$.

Let $B(\sigma(A), E)$ denote the space of all bounded E-measurable complex-valued functions on $\sigma(A)$. The integral of a function $f \in B(\sigma(A), E)$ is defined as follows. For $f = \sum_{j=1}^{n} \alpha_j \chi_{\varsigma_j}$ a finite simple function with $\alpha_j \in \mathbb{C}$ and $\varsigma_j \subseteq \sigma(A)$ mutually disjoint Borel sets for $j = 1, \ldots, n$, we let

$$\int_{\sigma(A)} f \, dE := \sum_{j=1}^{n} \alpha_j E(\varsigma_j).$$

This definition is independent of the representation of f, and

$$\left\| \int_{\sigma(A)} f \, dE \right\|_{\mathcal{X} \to \mathcal{X}} \leqslant 4 \, \mathrm{spec}(A) \|f\|_{\infty}.$$

Since the simple functions are dense in $B(\sigma(A), E)$, for a general $f \in B(\sigma(A), E)$ set

$$\int_{\sigma(A)} f \, dE := \lim_{n \to \infty} \int_{\sigma(A)} f_n \, dE \in \mathcal{B}(\mathcal{X}),$$

where $\{f_n\}_{n=1}^{\infty} \subseteq B(\sigma(A), E)$ is a sequence of simple functions satisfying $\|f_n - f\|_{\infty} \to 0$ as $n \to \infty$. This definition is independent of the choice of approximating sequence and

$$\left\| \int_{\sigma(A)} f\, dE \right\|_{X \to X} \leqslant 4 \operatorname{spec}(A) \|f\|_{\infty}. \tag{2.10.2}$$

It is straightforward to check that

$$\int_{\sigma(A)} (\alpha f + g)\, dE = \alpha \int_{\sigma(A)} f\, dE + \int_{\sigma(A)} g\, dE,$$

$$\int_{\sigma(A)} fg\, dE = \left(\int_{\sigma(A)} f\, dE \right) \left(\int_{\sigma(A)} g\, dE \right)$$

for all $\alpha \in \mathbb{C}$ and simple $f, g \in B(\sigma(A), E)$, and approximation then extends these identities to general $f, g \in B(\sigma(A), E)$. Moreover, $\int_{\sigma(A)} \chi_{\sigma(A)}\, dE = E(\sigma(A)) = I_X$. Hence the map $f \mapsto \int_{\sigma(A)} f\, dE$ is a continuous morphism $B(\sigma(A), E) \to \mathcal{B}(X)$ of unital Banach algebras. Since the spectrum of A is compact, the identity function is bounded on $\sigma(A)$ and $\int_{\sigma(A)} \lambda\, dE(\lambda) \in \mathcal{B}(X)$ is well defined.

Definition 2.10.1   A spectral operator $A \in \mathcal{B}(X)$ with spectral measure E is a scalar type operator if

$$A = \int_{\sigma(A)} \lambda\, dE(\lambda).$$

The class of scalar type operators on X is denoted by $\mathcal{B}_s(X)$.

For $A \in \mathcal{B}_s(X)$ with spectral measure E and $f \in B(\sigma(A), E)$ we define

$$f(A) := \int_{\sigma(A)} f\, dE.$$

As remarked above, $f \mapsto f(A)$ is a continuous morphism $B(\sigma(A), E) \to \mathcal{B}(X)$ of unital Banach algebras with norm bounded by $4 \operatorname{spec}(A)$.

Finally, we note that a normal operator A on a Hilbert space $\mathcal{H}$ is a scalar type operator with $\operatorname{spec}(A) = 1$, and in this case (2.10.2) improves to

$$\left\| \int_{\sigma(A)} f\, dE \right\| \leqslant \|f\|_{\infty},$$

as is known from the Borel functional calculus for normal operators.

Let now X possess an unconditional Schauder basis $\{e_j\}_{j=1}^{\infty} \subseteq X$ (see, e.g., [121]). For $j \in \mathbb{N}$, let $P_j \in \mathcal{B}(X)$ be the projection given by $P_j(x) = x_j e_j$ for all $x = \sum_{k=1}^{\infty} x_k e_k \in X$. An operator $A \in \mathcal{B}(X)$ is called diagonalizable (with respect to $\{e_j\}_{j=1}^{\infty})$ if there exist invertible $U \in \mathcal{B}(X)$ and a sequence $\{\lambda_j\}_{j=1}^{\infty} \in \ell^{\infty}$ of complex numbers such that

$$UAU^{-1}x = \sum_{j=1}^{\infty} \lambda_j P_j x, \quad x \in X,$$

where the series converges since $\{e_k\}_{k=1}^{\infty}$ is unconditional.

Observe that any diagonalizable operator is a scalar type operator.

2.11 Differentiability of Maps on Banach Spaces

The first order Fréchet and Gâteaux differentiability as well as higher order Fréchet differentiability are standard concepts (see, e.g., [181, Chapter I, Sections B and F]).

A map F between two Banach spaces $(X, \|\cdot\|_X)$ and $(Y, \|\cdot\|_Y)$ is called Fréchet differentiable at a point $x \in X$ if there exists $DF(x) \in \mathcal{B}(X, Y)$ such that for every $\varepsilon > 0$ there is $\delta > 0$, so that

$$\|F(x+h) - F(x) - DF(x)(h)\|_Y \leqslant \varepsilon \|h\|_X$$

for every $h \in X$ with $\|h\|_X < \delta$.

Note that if the map F is a bounded linear operator, then obviously $DF(x)(h) = F(h)$, that is, $DF(x) = F$ for every $x \in X$.

To define the higher order differentiability of a map between Banach spaces, we recall that there is an isometric isomorphism between the spaces $\mathcal{B}_2(X \times X, Y)$ and $\mathcal{B}(X, \mathcal{B}(X, Y))$; more generally, between $\mathcal{B}_n(X^{\times n}, Y)$ and $\mathcal{B}(X, \mathcal{B}(X, \ldots, \mathcal{B}(X, Y))\ldots)$. The latter identification of spaces is a standard fact of functional analysis (see, e.g., [181, Lemma 1.41]).

The map F is called twice Fréchet differentiable at a point $x \in X$ if it is Fréchet differentiable in a neighborhood of x and the respective differential $DF(x) \in \mathcal{B}(X, Y)$ is Fréchet differentiable at x. In this case, the second Fréchet differential $D^2F(x)$ is an element of $\mathcal{B}(X, \mathcal{B}(X, Y)) = \mathcal{B}_2(X \times X, Y)$. More generally, we have the following definition.

Definition 2.11.1 Let $n \in \mathbb{N}$. The map F between Banach spaces $(X, \|\cdot\|_X)$ and $(Y, \|\cdot\|_Y)$ is said to be n times Fréchet differentiable at $x \in X$ if it is $n-1$ times Fréchet differentiable in a neighborhood of x and there is $D^nF(x) \in \mathcal{B}_n(X^{\times n}, Y)$ satisfying

$$\left\|(D^{n-1}F(x+h) - D^{n-1}F(x))(h_1, \ldots, h_{n-1}) - D^nF(x)(h_1, \ldots, h_{n-1}, h)\right\|_Y$$

$$= o(\|h\|_X)\|h_1\|_X \cdots \|h_{n-1}\|_X$$

as $\|h\|_X \to 0$ for all $h_1, \ldots, h_{n-1} \in X$.

We further say that f is n times continuously Fréchet differentiable at x if it is n times Fréchet differentiable in a neighborhood of x and

$$\left\| (D^n F(x+h) - D^n F(x))(h_1, \ldots, h_n) \right\|_X = o(1) \|h_1\|_X \cdots \|h_n\|_X$$

as $\|h\|_X \to 0$ for all $h_1, \ldots, h_n \in X$.

Another definition of the higher order Fréchet differentiability is given in [126, Subsection 8.62]. To distinguish it from the differentiability defined above, we call it Taylor-Fréchet differentiability.

Definition 2.11.2 The map F between Banach spaces $(X, \| \cdot \|_X)$ and $(\mathcal{Y}, \| \cdot \|_{\mathcal{Y}})$ is called n-times Taylor-Fréchet differentiable at $x \in X$ if there exist $F_x^{(k)} \in \mathcal{B}_k(X^{\times k}, \mathcal{Y})$, $k = 1, \ldots, n$, such that for every $\varepsilon > 0$ there is $\delta > 0$ so that for every $h \in X$ with $\|h\|_X < \delta$,

$$\left\| F(x+h) - F(x) - \sum_{k=1}^n \frac{1}{k!} F_x^{(k)}(h, \ldots, h) \right\|_{\mathcal{Y}} \leqslant \varepsilon \|h\|_X^n.$$

It is proved in the proposition on pages 311–312 of [126] that Fréchet differentiability in the sense of Definition 2.11.1 implies the Taylor-Fréchet differentiability in the sense of Definition 2.11.2 and the respective equality $D^n F(x) = F_x^{(n)}$.

The map F between Banach spaces $(X, \| \cdot \|_X)$ and $(\mathcal{Y}, \| \cdot \|_{\mathcal{Y}})$ is called Gâteaux differentiable at x if it is differentiable along every direction in X, that is, there exists a map $D_G F(x)$ from X to $\mathcal{Y}$ defined as

$$D_G F(x)(h) := \| \cdot \|_{\mathcal{Y}^-} \lim_{t \to 0} \frac{F(x+th) - F(x)}{t} \tag{2.11.1}$$

for every $h \in X$. We note that $D_G F(x)$ is assumed to be neither bounded nor linear, but it will possess such properties in the problems that we consider.

The Fréchet differentiability implies the Gâteaux differentiability and equality

$$DF(x) = D_G F(x) \tag{2.11.2}$$

of the Fréchet and Gâteaux derivatives, but the converse holds only under an additional assumption on F. The following property is standard (see, e.g., [181, Lemma 1.15]).

Proposition 2.11.3 *Let F be a map between Banach spaces $(X, \| \cdot \|_X)$ and $(\mathcal{Y}, \| \cdot \|_{\mathcal{Y}})$. If F is Gâteaux differentiable in a neighborhood U of $x \in X$, $D_G F(\xi)$ given by (2.11.1) is an element of $\mathcal{B}(X, \mathcal{Y})$ for every $\xi \in U$, and $\xi \mapsto D_G F(\xi)$ is continuous at $\xi = x$, then F is Fréchet differentiable at x.*

We will understand higher order Gâteaux differentiability in the following sense.

Definition 2.11.4 Let $n \in \mathbb{N}$. The map F between Banach spaces $(\mathcal{X}, \| \cdot \|_{\mathcal{X}})$ and $(\mathcal{Y}, \| \cdot \|_{\mathcal{Y}})$ is said to be n times Gâteaux differentiable at $x \in \mathcal{X}$ if it is $n - 1$ times Gâteaux differentiable in a neighborhood of x and there is a map $D_G^n F(x)$ between Banach spaces $(\mathcal{X}, \| \cdot \|_{\mathcal{X}})$ and $(\mathcal{Y}, \| \cdot \|_{\mathcal{Y}})$ satisfying

$$\| \cdot \|_{\mathcal{Y}} \text{-} \lim_{t \to 0} \frac{D_G^{n-1} F(x + th)(h) - D_G^{n-1} F(x)(h)}{t} = D_G^n F(x)(h)$$

for every $h \in \mathcal{X}$.

In applications to differentiation of operator functions considered in Sect. 5.3, the map $D_G^n F(x)$ turns out to be bounded and homogeneous of order n.

When we discuss differentiability of operator functions we deal with modified concepts of Gâteaux and Fréchet derivatives. Such derivatives can be defined at points that do not belong to a Banach space, but are self-adjoint operators densely defined in a separable Hilbert space $\mathcal{H}$. Another modification consists in that the operator derivative can be calculated at an element of the Banach space $\mathcal{B}(\mathcal{H})$ while the direction of differentiation can be an element of a different Banach space, for instance, the Schatten ideal $\mathcal{S}^p$ or the noncommutative L^p-space $L^p(\mathcal{M}, \tau)$. To indicate that the operator derivative is calculated in the $\mathcal{S}^p$-, $\mathcal{L}^p(\mathcal{M}, \tau)$- or $L^p(\mathcal{M}, \tau)$-norm, we introduce the notations $D_{G,p}^n F(x)$ and $D_p^n F(x)$ for the respective nth order Gâteaux derivative and Fréchet differential of a function F at a point x along a direction in $\mathcal{S}^p$, $\mathcal{L}^p$, or L^p. If F is defined at self-adjoint operators, then directions of differentiation are automatically assumed to be self-adjoint.

Chapter 3
Double Operator Integrals

The concept of a double operator integral on $\mathcal{B}(\mathcal{H})$ was first introduced by
Yu. L. Daletskii and S. G. Krein in [61]. They launched this theory in order to
compute the derivative of the function $t \mapsto f(A(t))$, where $\{A(t)\}_t$ is a family of
bounded self-adjoint operators depending on the parameter t. Although the initial
construction allowed to handle a limited class of functions f and produced bounds
that depended on the spectra of operators, it created new conceptual and technical
opportunities. Further development of perturbation theory and its applications
stimulated extension and refinement of double operator integral constructions and
methods, with ground breaking contributions made in [34–36, 49, 141, 159]. There
are also generalizations of double operator integrals to multilinear transformations,
which are considered in the next section. In this chapter we discuss the main
constructions and properties of double operator integrals that have found important
applications.

3.1 Double Operator Integrals on Finite Matrices

Let $A, B \in \mathcal{B}_{sa}(\ell_d^2)$. Let $\{\xi_j\}_{j=1}^d$, $\{\eta_k\}_{k=1}^d$ be complete systems of orthonormal
eigenvectors and $\{\lambda_j\}_{j=1}^d, \{\mu_k\}_{k=1}^d$ sequences of the eigenvalues of A and B,
respectively. This notation is assumed throughout the section.

© Springer Nature Switzerland AG 2019
A. Skripka, A. Tomskova, *Multilinear Operator Integrals*,
Lecture Notes in Mathematics 2250, https://doi.org/10.1007/978-3-030-32406-3_3

3.1.1 Definition

Self-adjoint Case

Let $\varphi : \mathbb{R}^2 \to \mathbb{C}$. We define $T_\varphi^{A,B} : \mathcal{B}(\ell_d^2) \to \mathcal{B}(\ell_d^2)$ by

$$T_\varphi^{A,B}(X) := \sum_{j=1}^d \sum_{k=1}^d \varphi(\lambda_j, \mu_k) P_{\xi_j} X P_{\eta_k}, \qquad (3.1.1)$$

for $X \in \mathcal{B}(\ell_d^2)$. The operator $T_\varphi^{A,B}$ is a discrete version of a *double operator integral*. The function φ is usually called *the symbol* of the operator $T_\varphi^{A,B}$.

Let $\{\lambda_j\}_{j=1}^{d_0}$ and $\{\mu_k\}_{k=1}^{d_1}$ denote the sets of distinct eigenvalues of A and B, respectively. By properties of the spectral measure,

$$E_A(\{\lambda_i\}) = \sum_{\substack{1 \leqslant k \leqslant d \\ \lambda_k = \lambda_i}} P_{\xi_k}, \, i = 1, \ldots, d_0, \quad E_B(\{\mu_j\}) = \sum_{\substack{1 \leqslant k \leqslant d \\ \mu_k = \mu_j}} P_{\eta_k}, \quad j = 1, \ldots, d_1.$$

Thus, (3.1.1) can be rewritten in terms of the spectral measures of A and B as

$$T_\varphi^{A,B}(X) = \sum_{j=1}^{d_0} \sum_{k=1}^{d_1} \varphi(\lambda_j, \mu_k) E_A(\{\lambda_j\}) X E_B(\{\mu_k\}). \qquad (3.1.2)$$

Unitary Case

In the case when A, B are unitary operators the definition (3.1.1) is similar. The only difference is that the function φ is defined on the torus $\mathbb{T}^2$. Further in this section we distinguish the self-adjoint and unitary cases only if it is stated explicitly.

3.1.2 Relation to Finite-Dimensional Schur Multipliers

Let $c \in \ell_d^2$ have the representation $c = \sum_{l=1}^d c_l \, \eta_l$. Then the action of the operator $T_\varphi^{A,B}(X)$ on the element $c \in \ell_d^2$ can be written as follows:

$$T_\varphi^{A,B}(X)c = \sum_{j=1}^d \sum_{k=1}^d \sum_{l=1}^d c_l \, \varphi(\lambda_j, \mu_k) P_{\xi_j} X \langle \eta_l, \eta_k \rangle \eta_k$$

$$= \sum_{j=1}^d \sum_{k=1}^d c_k \, \varphi(\lambda_j, \mu_k) \langle X \eta_k, \xi_j \rangle \xi_j. \qquad (3.1.3)$$

Every linear operator $X \in \mathcal{B}(\ell_d^2)$ can be identified with the $d \times d$ complex matrix $X = \left(x_{jk}\right)_{j,k=1}^{d}$, where

$$x_{jk} = \langle X\eta_k, \xi_j \rangle.$$

Note that, since the systems $\{\xi_j\}_{j=1}^{d}$, $\{\eta_k\}_{k=1}^{d}$ are orthonormal, the (k_0, j_0)-entry for $1 \leqslant j_0, k_0 \leqslant d$ of the matrix corresponding to the operator $T_\varphi^{A,B}(X)$ given by (3.1.1) can be calculated from (3.1.3) as follows:

$$\langle T_\varphi^{A,B}(X)\eta_{k_0}, \xi_{j_0} \rangle = \sum_{j=1}^{d} \varphi(\lambda_j, \mu_{k_0})\langle\langle X\eta_{k_0}, \xi_j \rangle \xi_j, \xi_{j_0} \rangle$$

$$= \varphi(\lambda_{j_0}, \mu_{k_0})\langle X\eta_{k_0}, \xi_{j_0} \rangle = \varphi(\lambda_{j_0}, \mu_{k_0})x_{j_0 k_0}. \qquad (3.1.4)$$

Thus, the matrix $T_\varphi^{A,B}(X)$ is simply the matrix $\left(\varphi(\lambda_j, \mu_k)x_{jk}\right)_{j,k=1}^{d}$, that is, an entrywise product (usually called *a Schur product* or Hadamard product) of matrices $\left(\varphi(\lambda_j, \mu_k)\right)_{j,k=1}^{d}$ and $\left(x_{jk}\right)_{j,k=1}^{d}$. In other words, $T_\varphi^{A,B}$ acts as a *Schur multiplier* on $\mathcal{B}(\ell_d^2)$.

For other properties of finite dimensional Schur matrices we refer the reader to [32, 90]. Infinite dimensional Schur multipliers and their generalizations are discussed in [153, 203].

3.1.3   Properties of Finite Dimensional Double Operator Integrals

Algebraic Properties

The following properties of the operator integral $T_\varphi^{A,B}$ are direct consequences of the Schur multiplication properties and the representation (3.1.4) (for brevity, below we use the notation T_φ):

$$T_{\alpha\varphi+\beta\psi} = \alpha T_\varphi + \beta T_\psi, \quad \alpha, \beta \in \mathbb{C},$$

$$T_{\varphi\psi} = T_\varphi T_\psi,$$

$$T_\varphi = I, \quad \text{provided} \quad \varphi \equiv 1.$$

For future use, we make the next simple observation. Assume that the function φ has the representation

$$\varphi(\lambda, \mu) = a_1(\lambda)a_2(\mu)$$

for some complex valued functions a_1 and a_2. The definition of the double operator integral (3.1.1) and the spectral resolution of operator functions $a_1(A)$ and $a_2(B)$ imply that

$$T_{a_1 \cdot a_2}^{A,B}(X) = a_1(A) \cdot X \cdot a_2(B), \quad X \in \mathcal{B}(\ell_d^2). \tag{3.1.5}$$

The formula (3.1.5) lies at the foundation of the definition of a continuous variant of an operator integral (see Sect. 3.3).

Norm Estimates

An estimate for the Hilbert–Schmidt norm of $T_\varphi^{A,B}(X)$ is easy to obtain, as we see below.

Proposition 3.1.1 *Let* $A, B \in \mathcal{B}(\ell_d^2)$ *be self-adjoint (or unitary) operators with corresponding d-tuples of the eigenvalues* $\{\lambda_j\}_{j=1}^d$ *and* $\{\mu_k\}_{k=1}^d$. *Then, for a function* $\varphi : \mathbb{R}^2 \to \mathbb{C}$ *(or* $\varphi : \mathbb{T}^2 \to \mathbb{C}$*),*

$$\|T_\varphi^{A,B}(X)\|_2 \leqslant \max_{j,k} |\varphi(\lambda_j, \mu_k)| \, \|X\|_2. \tag{3.1.6}$$

Proof Combining (3.1.4) and (2.4.2) gives

$$\left\|T_\varphi^{A,B}(X)\right\|_2^2 = \sum_{j,k=1}^d |\varphi(\lambda_j, \mu_k)|^2 |x_{jk}|^2 \leqslant \max_{j,k} |\varphi(\lambda_j, \mu_k)|^2 \, \|X\|_2^2,$$

completing the proof. $\square$

The estimate (3.1.6) has a straightforward dimension-dependent analog (3.1.7) for the operator norm. More sophisticated, dimension-independent estimates for the operator norm of a double operator integral are discussed in Sect. 3.3.

Corollary 3.1.2 *Let* $A, B \in \mathcal{B}(\ell_d^2)$ *be self-adjoint (or unitary) operators with corresponding d-tuples of the eigenvalues* $\{\lambda_j\}_{j=1}^d$ *and* $\{\mu_k\}_{k=1}^d$. *Then, for a function* $\varphi : \mathbb{R}^2 \to \mathbb{C}$ *(or* $\varphi : \mathbb{T}^2 \to \mathbb{C}$*),*

$$\|T_\varphi^{A,B}(X)\| \leqslant \sqrt{d} \, \max_{j,k} |\varphi(\lambda_j, \mu_k)| \, \|X\|. \tag{3.1.7}$$

Proof By standard connections between different matrix norms (see, e.g., [32, Chapter IV, Section 2]),

$$\|T_\varphi^{A,B}(X)\| \leqslant \|T_\varphi^{A,B}(X)\|_2 \quad \text{and} \quad \|X\|_2 \leqslant \sqrt{d} \, \|X\|,$$

which along with (3.1.6) proves the result. $\square$

An estimate of type (3.1.6) is the best possible. We will strive to obtain analogous estimates for double (multiple) operator integrals for other norms on $\mathcal{B}(\ell_d^2)$ and its infinite dimensional analogs.

Perturbation Formula

The discrete symbol $f^{[1]}$ and the corresponding double operator integral was first studied by K. Löwner in [125].

Proposition 3.1.3 *Let* $A, B \in \mathcal{B}_{sa}(\ell_d^2)$ *and* $\sigma(A) \cup \sigma(B) \subset [a, b]$. *Let* $f : [a, b] \to \mathbb{C}$ *and let* $\psi : [a, b] \times [a, b] \to \mathbb{R}$ *be any function such that* $\psi(\lambda, \mu) = f^{[1]}(\lambda, \mu)$ *for all* $\lambda \in \sigma(A), \mu \in \sigma(B), \lambda \neq \mu$. *Then,*

$$f(A) - f(B) = T_\psi^{A,B}(A - B). \tag{3.1.8}$$

Proof By the spectral theorem,

$$A = \sum_{j=1}^d \lambda_j P_{\xi_j}, \quad B = \sum_{k=1}^d \mu_k P_{\eta_k}$$

and

$$f(A) = \sum_{j=1}^d f(\lambda_j) P_{\xi_j}, \quad f(B) = \sum_{k=1}^d f(\mu_k) P_{\eta_k}.$$

Since

$$f(A)\xi_j = f(\lambda_j)\xi_j, \quad f(B)\eta_k = f(\mu_k)\eta_k,$$

we have

$$\langle (f(A) - f(B))\eta_k, \xi_j \rangle = \psi(\lambda_j, \mu_k) \langle (A - B)\eta_k, \xi_j \rangle, \tag{3.1.9}$$

Rewriting Löwner's formula (3.1.9) in terms of the double operator integral and using (3.1.4), we obtain (3.1.8). □

A completely analogous result holds in the case of unitary operators.

Proposition 3.1.4 *Let* $U, V \in \mathcal{B}(\ell_d^2)$ *be unitary operators. If* f *is differentiable on* $\mathbb{T}$, *then*

$$f(U) - f(V) = T_{f^{[1]}}^{U,V}(U - V). \tag{3.1.10}$$

Lipschitz Estimates

The Lipschitz estimate for the Hilbert–Schmidt norm is a well-known result and is given in the following proposition. Similar estimate in the infinite-dimensional case is established in [36] and discussed in Sect. 5.1.1 (see Theorem 5.1.3).

Proposition 3.1.5 *Let $A, B \in \mathcal{B}_{sa}(\ell_d^2)$ and $\sigma(A) \cup \sigma(B) \subset [a, b]$. If $f \in \mathrm{Lip}[a, b]$, then*

$$\|f(A) - f(B)\|_2 \leqslant \|f\|_{\mathrm{Lip}[a,b]} \|A - B\|_2. \tag{3.1.11}$$

Proof Combining (3.1.8), (3.1.6), and (2.2.4) yields the result.   □

The estimate above is given in the case of self-adjoint operators A, B; however, similar argument shows that it also works for unitary operators with the constant $\|f'\|_\infty$.

The estimate (3.1.11) is the most desirable Lipschitz estimate. Estimates in other Schatten norms and in the operator norm require much more care since a straightforward application of (3.1.7), (2.2.4), and (3.1.8) gives the dimension-dependent estimate

$$\|f(A) - f(B)\| \leqslant \sqrt{d} \, \|f\|_{\mathrm{Lip}[a,b]} \|A - B\|. \tag{3.1.12}$$

The following result due to [8, Theorem 11.4] shows that the dimension-dependent component $\sqrt{d}$ in the inequality above can be improved.

Theorem 3.1.6 *Let $A, B \in \mathcal{B}_{sa}(\ell_d^2)$ and $\sigma(A) \cup \sigma(B) \subset [a, b]$. If $f \in \mathrm{Lip}[a, b]$, then*

$$\|f(A) - f(B)\| \leqslant C(1 + \log d)\|f\|_{\mathrm{Lip}[a,b]} \|A - B\|,$$

for some absolute constant C.

We will come back to this question in Sect. 5.1, where we study the Lipschitz estimate problem in the infinite-dimensional case. We will see, in particular, that in order to make the bounds for the operator and trace class norms of $f(A) - f(B)$ in terms of the same norm of $A - B$ independent of the dimension, we have to restrict the class of functions f.

Continuity

In the next lemma we state continuity of a double operator integral with respect to auxiliary self-adjoint (or unitary) matrix parameters. The proof is postponed to Proposition 4.1.6 applicable in the higher order case.

Proposition 3.1.7 *Let A_m, A, B_m, $B \in \mathcal{B}(\ell_d^2)$, $m \in \mathbb{N}$ be self-adjoint (respectively, unitary) operators such that $A_m \to A$ and $B_m \to B$ as $m \to \infty$. Let $\psi \in C_b(\mathbb{R}^2)$ (respectively, $\psi \in C(\mathbb{T}^2)$). Then,*

$$T_\psi^{A_m, B_m}(X) \to T_\psi^{A, B}(X), \quad m \to \infty,$$

for all $X \in \mathcal{B}(\ell_d^2)$.

3.2 Double Operator Integrals on $\mathcal{S}^2$

There are several approaches to double operator integrals on $\mathcal{S}^2$. In this section we define the continuous version of the double operator integral (3.1.1) as an integral with respect to the spectral measure constructed as a product of two other spectral measures. Such an approach was firstly suggested by M. S. Birman and M. Z. Solomyak in [34] (see also [35, 36]), where the authors have built the foundation of the theory of double operator integrals. We will also discuss an alternative approach due to B. S. Pavlov [138].

3.2.1 Definition

We start with the construction due to M. S. Birman and M. Z. Solomyak.

Birman–Solomyak's Approach

Let $\Omega = \Omega_1 \times \Omega_2$ and $\varphi \in L^\infty(\Omega, G)$, where G is defined in (2.5.1). Recall that $L^\infty(\Omega, G)$ consists of all G-measurable functions φ on Ω for which

$$\|\varphi\|_\infty := G\text{-}\sup \varphi = \inf\{\alpha \in \mathbb{R}_+ : |\varphi(\cdot)| \leqslant \alpha, \, G\text{-a.e.}\}$$

Definition 3.2.1 The Birman–Solomyak double operator integral $T_\varphi^G : \mathcal{S}^2 \to \mathcal{S}^2$ is defined as the integral of the symbol φ with respect to the spectral measure G, that is,

$$T_\varphi^G(X) := \int_\Omega \varphi(\omega) \, dG(\omega)(X), \quad X \in \mathcal{S}^2. \tag{3.2.1}$$

The notation

$$T_\varphi^G(X) =: \int_{\Omega_1 \times \Omega_2} \varphi(\omega_1, \omega_2) dE(\omega_1) X dF(\omega_2), \quad X \in \mathcal{S}^2,$$

is frequently used.

The property $\| \int_\Omega \varphi(\omega) \, dG(\omega) \| = \| \varphi \|_{L^\infty(\Omega, G)}$ of the functional calculus implies the following important result.

Proposition 3.2.2 *The operator integral* T_φ^G *is bounded on* $\mathcal{S}^2$ *if and only if* $\varphi \in L^\infty(\Omega, G)$. *In this case,*

$$\| T_\varphi^G : \mathcal{S}^2 \to \mathcal{S}^2 \| = \| \varphi \|_{L^\infty(\Omega, G)}. \tag{3.2.2}$$

Let $A, B \in \mathcal{D}_{sa}$ and let E and F be spectral measures of A and B, respectively. Let $\Omega_1 = \Omega_2 = \mathbb{R}$ and let $\Sigma_1 = \Sigma_2$ be the σ-algebra of all Borel sets on $\mathbb{R}$. In this case, we use the notation

$$T_\varphi^{A,B} := T_\varphi^G, \tag{3.2.3}$$

for $\varphi \in L^\infty(\mathbb{R}^2, G)$.

Pavlov's Approach

If $Y \in \mathcal{S}^2$ and $\varphi \in L^\infty(\Omega, G)$, where G is given by (2.5.1), then $\sigma \mapsto \text{Tr}(G(\sigma)Y)$ is a scalar measure on $\Sigma = \Sigma_1 \otimes \Sigma_2$ and φ is integrable with respect to $\text{Tr}(G(\cdot)Y)$ (for the construction of G see Proposition 2.5.1). Thus,

$$\theta_G : Y \mapsto \int_\Omega \varphi(\omega_1, \omega_2) \, d \, \text{Tr}(G(\omega_1 \times \omega_2)Y) \tag{3.2.4}$$

is a linear bounded functional on $\mathcal{S}^2$. Since $(\mathcal{S}^2)^* = \mathcal{S}^2$, it follows that there exists $A_G \in \mathcal{S}^2$ such that

$$\theta_G(Y) = \text{Tr}(A_G Y)$$

for all $Y \in \mathcal{S}^2$. We set

$$\int_\Omega \varphi(\omega_1, \omega_2) dG(\omega_1 \times \omega_2) := A_G.$$

The transformation A_G is the Pavlov's version of the double operator integral. Applying the double operator integral (3.2.1) to $Y \in \mathcal{S}^2$ and taking the trace, we obtain $\theta_G(Y)$, where θ_G is defined in (3.2.4). Therefore, Pavlov's definition and Birman–Solomyak's one coincide (see also [34, (2.12)]).

We will discuss Pavlov's approach [138] in full generality of arbitrary order multiple operator integrals in Sect. 4.2.

3.2.2 Relation to Schur Multipliers on $\mathcal{S}^2$

Fix two orthonormal bases $\{\xi_j\}_{j=1}^\infty$ and $\{\eta_k\}_{k=1}^\infty$ in $\mathcal{H}$. Then, every operator $X \in \mathcal{S}^2$ can be represented as an infinite matrix $X = (x_{jk})_{j,k=1}^\infty$, where $x_{jk} = \langle X(\eta_k), \xi_j \rangle$, $j, k \in \mathbb{N}$. For a matrix $Y = (y_{jk})_{j,k=1}^\infty$ let $X \circ Y := (y_{jk}x_{jk})_{j,k=1}^\infty$ denote the Schur product of the matrices X and Y. The matrix Y is called a *Schur multiplier* if the mapping $X \mapsto Y \circ X$ is a bounded operator on $\mathcal{S}^2$. We denote by $\mathcal{M}(\mathcal{S}^2, \mathcal{S}^2)$ the space of all multipliers on $\mathcal{S}^2$ with the norm

$$\|Y\|_{2,2} := \sup\{\|Y \circ X\|_2 \ : \ \|X\|_2 \leqslant 1\}.$$

Suppose that each of $A, B \in \mathcal{B}_{sa}(\mathcal{H})$ has a discrete spectrum. Consider sequences $\lambda = \{\lambda_j\}_{j=1}^\infty$ and $\mu = \{\mu_k\}_{k=1}^\infty$ consisting of the points of the spectrum of A and B, respectively, counted with multiplicities. Let $\{\xi_j\}_{j=1}^\infty$, $\{\eta_k\}_{k=1}^\infty$ be corresponding orthonormal bases of eigenvectors of the operators A and B, respectively, and let $\varphi \in L^\infty(\mathbb{R}^2, G)$. Then, we have the formulas

$$T_\varphi^{A,B}(X) = \sum_{j,k} \varphi(\lambda_j, \mu_k) P_{\xi_j} X P_{\eta_k},$$

$$\langle T_\varphi^{A,B}(X)\eta_k, \xi_j \rangle = \varphi(\lambda_j, \mu_k)x_{jk}.$$

Thus, $T_\varphi^{A,B}(X)$ is represented as the Schur product of the matrices $(\varphi(\lambda_j, \mu_k))_{j,k}$ and $X = (x_{jk})_{jk}$ (compare with finite versions (3.1.1) and (3.1.4)).

The double operator integral can be viewed as a continuous version of a Schur multiplier and, therefore, sometimes the operator integral $T_\varphi^{A,B}$ is called a Schur multiplier. Observe also that since the operator $T_\varphi^{A,B}$ is bounded on $\mathcal{S}^2$ if and only if $\varphi \in L^\infty(\mathbb{R}^2, G)$ (see Proposition 3.2.2), it follows that the matrix $Y = (y_{jk})_{jk}$ is the Schur multiplier on $\mathcal{S}^2$ if and only if $\sup_{j,k} |y_{jk}| < \infty$. In this case,

$$\|Y\|_{2,2} = \sup_{j,k} |y_{jk}|$$

(for an alternative proof of the latter result see, e.g., [17, Proposition 2.1]).

When A and B have arbitrary spectra, the transformation $T_\varphi^{A,B}$ can be viewed as a Schur multiplier acting by pointwise multiplication on direct integral decompositions of operators (see, e.g., [41, Subsection 3.2] or [203]).

3.2.3 Basic Properties of Double Operator Integrals on S^2

Algebraic Properties

The following properties of the operator integral $T_\varphi^{A,B} = T_\varphi^G$ on S^2 defined in (3.2.1) and (3.2.3) hold (for brevity, we use the notation T_φ):

$$T_{\alpha\varphi+\beta\psi} = \alpha T_\varphi + \beta T_\psi, \quad \alpha, \beta \in \mathbb{C};$$

$$T_{\varphi\psi} = T_\varphi T_\psi;$$

$$T_\varphi = I, \quad \text{provided} \quad \varphi \equiv 1;$$

$$T_{\overline{\varphi}} = T_\varphi^*;$$

where $\overline{\varphi}$ denotes the complex conjugate of φ.

Proposition 3.2.3 *Assume that $\varphi \in L^\infty(\Omega, G)$, where G is defined in (2.5.1), has a representation $\varphi(\omega_1, \omega_2) = a_1(\omega_1)a_2(\omega_2)$ for some $a_1, a_2 \in L^\infty(\Omega, G)$. Then, the formula*

$$T_\varphi^G(X) = \int_{\Omega_1} a_1(\omega_1)dE(\omega_1) \cdot X \cdot \int_{\Omega_2} a_2(\omega_2)dF(\omega_2) \tag{3.2.5}$$

holds for all $X \in S^2$.

Proof Indeed, observing that

$$T_{a_1}^G(X) = T_{a_1}^{\mathcal{E}}(X) = \int_{\Omega_1} a_1(\omega_1)d\mathcal{E}(\omega_1)(X) = \int_{\Omega_1} a_1(\omega_1)dE(\omega_1) \cdot X, \tag{3.2.6}$$

and, similarly,

$$T_{a_2}^G(X) = T_{a_2}^{\mathcal{F}}(X) = X \cdot \int_{\Omega_2} a_2(\omega_2)dF(\omega_2), \tag{3.2.7}$$

and using multiplicativity of the spectral integral, we obtain

$$T_\varphi^G(X) = T_{a_1 a_2}^G(X) = (T_{a_1}^G T_{a_2}^G)(X) = T_{a_1}^G \left(X \cdot \int_{\Omega_2} a_2(\omega_2)dF(\omega_2) \right)$$

$$= \int_{\Omega_1} a_1(\omega_1)dE(\omega_1) \cdot X \cdot \int_{\Omega_2} a_2(\omega_2)dF(\omega_2).$$

Corollary 3.2.4 Let $\varphi \in L^\infty(\mathbb{R}^2, G)$ have the representation $\varphi(\lambda, \mu) = a_1(\lambda)a_2(\mu)$ for some $a_1, a_2 \in L^\infty(\Omega, G)$. Then,

$$T_\varphi^G(X) = T_\varphi^{A,B}(X) = a_1(A) \cdot X \cdot a_2(B). \qquad (3.2.8)$$

Proof The result follows from (3.2.5) upon applying the spectral resolutions of $a_1(A)$ and $a_2(B)$. $\qquad \square$

The resolution (3.2.8) extends its finite-dimensional version (3.1.5); it plays a crucial role in the further presentation. In particular, one of definitions of the double operator integral on $\mathcal{B}(\mathcal{H})$ and Schatten ideals is based on this formula (see Sect. 3.3).

Further properties of double operator integrals on $\mathcal{S}^2$ are collected in Sect. 3.3.5.

3.3 Double Operator Integrals on Schatten Classes and $\mathcal{B}(\mathcal{H})$

We start with a brief discussion of the original construction due to Yu. L. Daletskii and S. G. Krein [61] and then turn to a detailed discussion of constructions frequently used in contemporary analysis.

3.3.1 Daletskii-Krein's Approach

The double operator integral in [61] was introduced as an iterated integral on $\mathcal{B}(\mathcal{H})$. We present the definition below.

Let $A \in \mathcal{B}_{sa}(\mathcal{H})$, let $[a, b]$ contain $\sigma(A)$, and let $\{E_\mu\}_{a \leqslant \mu \leqslant b}$ be the spectral family of A. Let $F(\mu)$ be a bounded operator on $\mathcal{H}$ depending on the parameter μ. Consider the abstract Stieltjes integral

$$\int_a^b F(\mu) \, dE_\mu, \qquad (3.3.1)$$

which is defined as the limit of the operators

$$F_\delta = \sum_{k=1}^n F(\mu_k) E(\delta_k)$$

in the operator norm as the maximal length of the intervals $\delta_k = [\lambda_{k-1}, \lambda_k]$ tends to zero, where $\delta = \{\delta_k\}$ is a partition of $[a, b]$ by the points $a = \lambda_0 < \lambda_1 < \cdots < \lambda_{n-1} < \lambda_n = b$ and $\mu_k \in \delta_k$. If the integral in (3.3.1) exists, it is called an operator integral. The following estimate is obtained in [61, Theorem 1.4].

Theorem 3.3.1 *Let $\varphi(\lambda, \mu)$ be a continuous operator function with partial derivatives $\varphi'_\lambda(\lambda, \mu)$, $\varphi'_\mu(\lambda, \mu)$, $\varphi''_{\lambda\mu}(\lambda, \mu)$ continuous in the rectangle $a \leqslant \lambda, \mu \leqslant b$ and let X be a bounded operator which depends on neither λ nor μ. Then the iterated integral*

$$T_\varphi^{A,A}(X) := \int_a^b \left[\int_a^b \varphi(\lambda, \mu)\, dE_\lambda \right] X\, dE_\mu$$

exists and

$$\left\| \int_a^b \left[\int_a^b \varphi(\lambda, \mu)\, dE_\lambda \right] X\, dE_\mu \right\|$$

$$\leqslant \|X\| \Big(\max_{\lambda,\mu} |\varphi(\lambda, \mu)| + (b - a) \max_{\lambda,\mu} |\varphi'_\lambda(\lambda, \mu)| \qquad (3.3.2)$$

$$+ (b - a) \max_{\lambda,\mu} |\varphi'_\mu(\lambda, \mu)| + (b - a)^2 \max_{\lambda,\mu} |\varphi''_{\lambda\mu}(\lambda, \mu)| \Big).$$

The operator $T_\varphi^{A,A}$ is the first version of the double operator integral on $\mathcal{B}(\mathcal{H})$ and if the Hilbert space $\mathcal{H}$ is finite-dimensional, then the definition of the double operator integral above coincides with (3.1.1).

The estimate (3.3.2) is the best one attained for the operator integral defined in [61], but it depends on the size of the spectrum of A (in particular, does not apply to unbounded A) and imposes unnecessary restrictions on the symbol φ. There exist other constructions of double operator integrals, which are presented further in the text, that provide better estimates under weaker restrictions on the operators and symbols.

3.3.2 Extension from the Double Operator Integral on $\mathcal{S}^2$

Another approach of double operator integral in $\mathcal{B}(\mathcal{H})$ is suggested by M. S. Birman and M. Z. Solomyak in [34] (see also [41, Section 4]), where the authors extend the double operator integral from $\mathcal{S}^2$ to $\mathcal{B}(\mathcal{H})$, with an additional assumption on the symbol φ.

Recall that

$$\mathcal{S}^1 \subset \mathcal{S}^2 \subset \mathcal{B}(\mathcal{H})$$

with continuous inclusions. Moreover,

$$(\mathcal{S}^1)^* = \mathcal{B}(\mathcal{H}).$$

For any L^∞-function φ, we have that the operator T_φ maps $\mathcal{S}^1$ into $\mathcal{S}^2$. If φ is such that the image $T_\varphi(\mathcal{S}^1)$ lies in $\mathcal{S}^1$ and the restriction $T_\varphi|_{\mathcal{S}^1}$ is a bounded operator on

S^1, then we say that the double operator integral T_φ is bounded on S^1. One can show that in this case the operator $T_{\overline{\varphi}}$ is also bounded on S^1 and has the same norm. Then the adjoint operator $T_{\overline{\varphi}}|_{S^1}^*$ acts on the space $\mathcal{B}(\mathcal{H})$. Hence, we naturally define

$$T_\varphi(X) := (T_{\overline{\varphi}}|_{S^1}^*)(X), \qquad X \in \mathcal{B}(\mathcal{H}).$$

If $X \in S^\infty$, then also $T_\varphi(X) \in S^\infty$. Indeed, for any finite rank operator X, we have $T_\varphi(X) \in S^1 \subset S^\infty$ and the set of all finite rank operators is dense in S^∞. So the operator T_φ defined above acts from S^∞ to S^∞. Moreover,

$$\|T_\varphi : S^\infty \to S^\infty\| = \|T_\varphi : \mathcal{B}(\mathcal{H}) \to \mathcal{B}(\mathcal{H})\| = \|T_\varphi : S^1 \to S^1\|$$

and

$$\|T_\varphi : \mathcal{B}(\mathcal{H}) \to \mathcal{B}(\mathcal{H})\| \geqslant \|T_\varphi : S^2 \to S^2\| = \|\varphi\|_\infty.$$

We discuss the existence of this double operator integral and its estimate later, in Sect. 3.3.6.

3.3.3   Approach via Separation of Variables

This approach was introduced by M. S. Birman and M. Z. Solomyak in [34, Section 6] and developed further [35, 36, 141]. It is now a particular case of the theory of multilinear operator integrals constructed in [24, 146].

Demonstration in a Simple Case

The simplest approach to the definition of a double operator integral on S^p is to use the property (3.2.8). Let $A, B \in \mathcal{B}_{sa}(\mathcal{H})$ be self-adjoint operators and let $\varphi \in L^\infty(\mathbb{R}^2)$ be given by the formula

$$\varphi(\lambda, \mu) = \sum_{j=1}^{n} a_1(j, \lambda) a_2(j, \mu), \quad \lambda, \mu \in \mathbb{R},$$

where $a_1(j, \cdot), a_2(j, \cdot) \in L^\infty(\mathbb{R})$, $j = 1, \ldots, n$. Define the operator $T_\varphi^{A,B} : S^p \to S^p$ by

$$T_\varphi^{A,B}(X) := \sum_{j=1}^{n} a_1(j, A) X a_2(j, B).$$

From the Hölder inequality (2.4.1) it follows that $T_\varphi^{A,B}$ is a bounded linear operator on $\mathcal{S}^p$ for $1 \leqslant p \leqslant \infty$ and

$$\|T_\varphi^{A,B} : \mathcal{S}^p \to \mathcal{S}^p\| \leqslant \sum_{j=1}^n \|a_1(j, \cdot)\|_\infty \|a_2(j, \cdot)\|_\infty.$$

Similarly, the operator $T_\varphi^{A,B}$ is defined on $\mathcal{B}(\mathcal{H})$.

General Case

Below we present the largest class of functions to which the above approach applies. This class is the integral projective tensor product of L^∞-spaces introduced in [141].

Let $\mathfrak{A}_1$ be the class of functions $\varphi : \mathbb{R}^2 \to \mathbb{C}$ admitting the representation

$$\varphi(\lambda, \mu) = \int_\Omega a_1(\lambda, \omega)\, a_2(\mu, \omega) dv(\omega), \qquad (3.3.3)$$

for some finite measure space (Ω, ν) and bounded measurable functions

$$a_i\, (\cdot, \cdot) : \mathbb{R} \times \Omega \to \mathbb{C}, \;\; i = 1, 2,$$

where on $\mathbb{R}$ we consider the Borel σ-algebra, such that

$$\int_\Omega \|a_1(\cdot, \omega)\|_\infty \|a_2(\cdot, \omega)\|_\infty \, d\,|\nu|\,(\omega) < \infty.$$

The class $\mathfrak{A}_1$ is an algebra with respect to the operations of pointwise addition and multiplication [24, Proposition 4.10]. The formula

$$\|\varphi\|_{\mathfrak{A}_1} = \inf \int_\Omega \|a_1(\cdot, \omega)\|_\infty \|a_2(\cdot, \omega)\|_\infty \, d\,|\nu|\,(\omega),$$

where the infimum is taken over all possible representations (3.3.3), defines a norm on $\mathfrak{A}_1$ (see [63, Lemma 4.6]).

The class $\mathfrak{A}_1$ can also be defined as the class of functions $\varphi : \mathbb{R}^2 \to \mathbb{C}$ admitting the representation

$$\varphi(\lambda, \mu) = \int_\Omega b_1(\lambda, \omega)\, b_2(\mu, \omega) \, dv_2(\omega), \qquad (3.3.4)$$

for some (not necessarily finite) measure space (Ω, ν_2) and bounded measurable functions

$$b_i\, (\cdot, \cdot) : \mathbb{R} \times \Omega \to \mathbb{C}, \;\; i = 1, 2,$$

such that

$$\int_{\Omega} \|b_1(\cdot, \omega)\|_{\infty} \|b_2(\cdot, \omega)\|_{\infty} \, d \, |v_2| \, (\omega) < \infty$$

(see, e.g., [63, 158]). These definitions coincide since the representation (3.3.3) of the function φ can be obtained from (3.3.4) with

$$a_1(\lambda, \omega) = \frac{b_1(\lambda, \omega)}{\|b_1(\cdot, \omega)\|_{\infty}}, \quad a_2(\lambda, \omega) = \frac{b_2(\lambda, \omega)}{\|b_2(\cdot, \omega)\|_{\infty}}$$

and the finite measure v defined by

$$v = \|b_1(\cdot, \omega)\|_{\infty} \|b_2(\cdot, \omega)\|_{\infty} v_2.$$

Another important description of the class $\mathfrak{A}_1$ is given in [141, Theorem 1, (3)$\Leftrightarrow$(4)] (for (3)$\Leftarrow$(4) see [36]). Namely, $\mathfrak{A}_1$ can also be defined as the class of functions $\varphi : \mathbb{R}^2 \to \mathbb{C}$ admitting the representation

$$\varphi(\lambda, \mu) = \int_{\Omega_1} c_1(\lambda, \omega) \, c_2(\mu, \omega) \, dv_1(\omega),$$

for some (not necessarily finite) measure space (Ω_1, v_1) and measurable functions

$$c_i(\cdot, \cdot) : \mathbb{R} \times \Omega_1 \to \mathbb{C}, \quad i = 1, 2,$$

such that

$$\left\| \int_{\Omega_1} |c_1(\cdot, \omega)|^2 dv_1(\omega) \right\|_{\infty} \left\| \int_{\Omega_1} |c_2(\cdot, \omega)|^2 dv_1(\omega) \right\|_{\infty} < \infty.$$

Definition

The definition of a double operator integral based on a separation of variables is given below.

Definition 3.3.2 Let $1 \leqslant p \leqslant \infty$. For every $\varphi \in \mathfrak{A}_1$, and a fixed couple $A, B \in \mathcal{D}_{sa}$, the double operator integral $T_{\varphi}^{A,B} : S^p \to S^p$ (respectively, $T_{\varphi}^{A,B} : \mathcal{B}(\mathcal{H}) \to \mathcal{B}(\mathcal{H})$) is defined by

$$T_{\varphi}^{A,B}(X) := \int_{\Omega} a_1(A, \omega) \, X \, a_2(B, \omega) \, dv(\omega),$$

for $X \in \mathcal{S}^p$ (respectively, $X \in \mathcal{B}(\mathcal{H})$), where a_j and (Ω, ν) are taken from the representation (3.3.3) and the integral is understood in the sense of the Bochner integral

$$\left(\int_\Omega a_1(A, \omega) X a_2(B, \omega) \, d\nu(\omega) \right)(y) = \int_\Omega \left(a_1(A, \omega) X a_2(B, \omega) \right)(y) \, d\nu(\omega), \quad y \in \mathcal{H}$$

(see, e.g., [141] and also [24, Definition 4.1]).

It follows directly from the definition that if $A, B \in \mathcal{D}_{sa}$ and $\varphi \in \mathfrak{A}_1$, then $T_\varphi^{A,B}$ is a bounded linear operator on $\mathcal{S}^p$, $1 \leqslant p \leqslant \infty$, and on $\mathcal{B}(\mathcal{H})$.

One of important results of this theory is that the value $T_\varphi^{A,B}(X)$ does not depend on the particular representation on the right-hand side of (3.3.3) (see [71, Lemma 7.2 and Theorem 7.5]).

Similarly to the proof of (3.2.8), one can show that if $\varphi \in \mathfrak{A}_1$, then $T_\varphi^{A,B}$ on $\mathcal{S}^2$ given by Definition 3.3.2 coincides with $T_\varphi^{A,B}$ given by Definition 3.2.1 of Sect. 3.2, and so these definitions also agree with the definition of double operator integral on $\mathcal{S}^1, \mathcal{S}^\infty$, and $\mathcal{B}(\mathcal{H})$ given in Sect. 3.3.2.

3.3.4 Approach Without Separation of Variables

An approach without separation of variables that enables strong estimates of Schatten norms of double operator integrals was introduced by D. Potapov and F. Sukochev in [159]. This double operator integral is a particular case of a general order multiple operator integral introduced in [163], which is discussed in Chap. 4. In the present subsection we follow [159] and consider only the special case of operators with spectra consisting of a finite number of rational points.

Let A and B be operators with spectra contained in $\left\{ -\frac{N}{m}, \ldots, \frac{N}{m} \right\}$ for some $m, N \in \mathbb{N}$ and let $\varphi : \mathbb{R}^2 \to \mathbb{C}$ be a bounded Borel function. The double operator integral associated with A, B, φ is defined on $\mathcal{B}(\mathcal{H})$ by

$$T_\varphi^{A,B}(X) := \sum_{|l_0|, |l_1| \leqslant N} \varphi\left(\frac{l_0}{m}, \frac{l_1}{m} \right) E_A\left(\left[\frac{l_0}{m}, \frac{l_0}{m} + 1 \right) \right) X E_B\left(\left[\frac{l_1}{m}, \frac{l_1}{m} + 1 \right) \right),$$

(3.3.5)

for $X \in \mathcal{B}(\mathcal{H})$.

In the case when A, B are finite dimensional matrices, the transformation defined in (3.3.5) coincides with the one in (3.1.2).

It follows from the spectral theorem that the transformations given by Definition 3.3.2 and by the formula (3.3.5) coincide for all $\varphi \in \mathfrak{A}_1$ and all A, B with spectra contained in $\left\{ -\frac{N}{m}, \ldots, \frac{N}{m} \right\}$ for some $m, N \in \mathbb{N}$.

3.3.5   Properties of Double Operator Integrals on $\mathcal{S}^p$ and $\mathcal{B}(\mathcal{H})$

Algebraic Properties

It is established in [24, Proposition 4.10] that the double operator integral given by Definition 3.3.2 satisfies

$$T_{\alpha_1\varphi_1+\alpha_2\varphi_2} = \alpha_1 T_{\varphi_1} + \alpha_2 T_{\varphi_2},$$
$$T_{\varphi_1\varphi_2} = T_{\varphi_1} T_{\varphi_2},$$

for $\varphi_1, \varphi_2 \in \mathfrak{A}_1, \alpha_1, \alpha_2 \in \mathbb{C}$.

Norm Estimates

The following estimate for a double operator integral is an immediate consequence of Definition 3.3.2 and the ideal property of Schatten classes.

Theorem 3.3.3 *Let $A, B \in \mathcal{D}_{sa}$, $\varphi \in \mathfrak{A}_1$, and $1 \leqslant p \leqslant \infty$. Then,*

$$\left\| T_\varphi^{A,B}(X) \right\|_p \leqslant \|\varphi\|_{\mathfrak{A}_1} \|X\|_p \tag{3.3.6}$$

for every $X \in \mathcal{S}^p$ if $1 \leqslant p < \infty$ or $X \in \mathcal{B}(\mathcal{H})$ if $p = \infty$.

The double operator integral (3.3.5) as well as its continuous version with symbol φ equal to the divided difference of a Lipschitz function discussed in Sect. 3.3.2 admits extension and improvement of the estimate (3.3.6) on the Schatten classes $\mathcal{S}^p$, $p > 1$. The following result is established in [159, Theorem 7].

Theorem 3.3.4 *Let $A, B \in \mathcal{D}_{sa}$, $f \in \mathrm{Lip}(\mathbb{R})$, and $1 < p < \infty$. Then, there exists $c_p > 0$ such that*

$$\|T_{f^{[1]}}^{A,B}(X)\|_p \leqslant c_p \|f\|_{\mathrm{Lip}(\mathbb{R})} \|X\|_p$$

for every $X \in \mathcal{S}^p$, where $f^{[1]}$ is defined in (2.2.1).

The original version of the proof of Theorem 3.3.4 in [159, Theorem 7] is based on decomposition of the symbol $f^{[1]}$ that realizes $T_{f^{[1]}}^{A,B}$ as a suitable combination of triangular truncations and multipliers on UMD spaces handled by Marcinkiewicz-Mihlin multiplier theory. Results on UMD spaces can be found in, for instance, [91]; in particular, the mentioned multiplier theory is applied in [91, Proposition 5.48] and discussed in references cited therein. In addition, the result of Theorem 3.3.4 is derived in the proof of the basis of induction for Theorem 4.3.10 with involvement of a vector-valued harmonic analysis and interpolation. It can also be derived from the boundedness of the double operator integral $\mathcal{S}^1 \to \mathcal{S}^{1,\infty}$ (discussed in Sect. 3.3.7) and the double operator integral $S^2 \to S^2$ for $1 < p < 2$ by interpolation and then extended to $2 < p < \infty$ by duality.

The sharp estimate for the constant c_p from Theorem 3.3.4 in terms of p is established in [48] based on harmonic analysis of UMD spaces S^p, $1 < p < \infty$.

Theorem 3.3.5 *The constant c_p from Theorem 3.3.4 satisfies*

$$c_p \sim \frac{p^2}{p-1}, \quad 1 < p < \infty.$$

Perturbation Formula

The following nice property of the double operator integral $T^{A,B}_{f^{[1]}}$ is obtained in [39, Theorem 2.1]. It also admits an extension to the case of unbounded operators A and B [39, Theorem 2.2].

Theorem 3.3.6 *Let $A, B \in \mathcal{B}_{sa}(\mathcal{H})$ with $\sigma(A) \cup \sigma(B) \subseteq [a, b]$ and $X \in S^2$. If $f \in \mathrm{Lip}[a, b]$, then*

$$f(A)X - Xf(B) = T^{A,B}_{f^{[1]}}(AX - XB). \qquad (3.3.7)$$

Proof Let $p_1(\lambda, \mu) = \lambda$ and $p_2(\lambda, \mu) = \mu$, for $\lambda, \mu \in \mathbb{R}$. Since f is Lipschitz, it follows that f is continuous on $[a, b]$ and, therefore, $f \circ p_1, f \circ p_2 \in L^\infty([a, b] \times [a, b], G)$, where G is defined in (2.5.1). Moreover, since $(f \circ p_1)(\lambda, \mu) = f(p_1(\lambda, \mu)) = f(\lambda)$ and $(f \circ p_2)(\lambda, \mu) = f(\mu)$, by (3.2.6) and (3.2.7), we have

$$T^{A,B}_{f \circ p_1}(X) = f(A)X, \quad T^{A,B}_{f \circ p_2}(X) = Xf(B),$$

$$T^{A,B}_{p_1}(X) = AX, \quad T^{A,B}_{p_2}(X) = XB,$$

for $X \in S^2$. Thus, using basic properties of double operator integrals, we obtain

$$f(A)X - Xf(B) = T^{A,B}_{f \circ p_1}(X) - T^{A,B}_{f \circ p_2}(X) = T^{A,B}_{f \circ p_1 - f \circ p_2}(X)$$

$$= T^{A,B}_{(p_1 - p_2)f^{[1]}}(X) = T^{A,B}_{p_1 f^{[1]}}(X) - T^{A,B}_{p_2 f^{[1]}}(X)$$

$$= T^{A,B}_{f^{[1]}}(T^{A,B}_{p_1}(X)) - T^{A,B}_{f^{[1]}}(T^{A,B}_{p_2}(X)) = T^{A,B}_{f^{[1]}}(AX - XB).$$

Although Theorem 3.3.6 is not applicable to the identity operator $X = I$, an analog of (3.1.8) remains valid. This result is the Consequence of [36, Theorem 4.5].

Theorem 3.3.7 *Let $A, B \in \mathcal{D}_{sa}$ be such that $A - B \in \mathcal{S}^2$ and let $f \in \mathrm{Lip}(\mathbb{R})$. Then, $f(A) - f(B) \in \mathcal{S}^2$ and*

$$f(A) - f(B) = T_{f^{[1]}}^{A,B}(A - B). \tag{3.3.8}$$

We omit the proof of Theorem 3.3.7, but demonstrate its major ideas later, in the proof of Theorem 4.4.8.

The above representation for an increment of an operator function extends to the case of non-Hilbert–Schmidt perturbations under an additional assumption on the scalar function. The next result is due to [36, Consequence of Theorem 4.5].

Theorem 3.3.8 *Let $A, B \in \mathcal{D}_{sa}$ be such that $A - B \in \mathcal{S}^p$, $1 \leqslant p \leqslant \infty$ ($\mathcal{B}(\mathcal{H})$, respectively). If $f \in \mathrm{Lip}(\mathbb{R})$ such that $f^{[1]} \in \mathfrak{A}_1$, then $f(A) - f(B) \in \mathcal{S}^p$ ($\mathcal{B}(\mathcal{H})$, respectively) and*

$$f(A) - f(B) = T_{f^{[1]}}^{A,B}(A - B). \tag{3.3.9}$$

Continuity

Let $\mathfrak{C}_1$ denote the subset of $\mathfrak{A}_1$ of functions admitting the representation (3.3.3), where $\cup_{k=1}^{\infty} \Omega_k = \Omega$ for a growing sequence $\{\Omega_k\}_{k=1}^{\infty}$ of measurable subsets of Ω such that the families $\{a_j(\cdot, \omega)\}_{\omega \in \Omega_k}$, $j = 1, 2$, are uniformly bounded and uniformly continuous. A norm on $\mathfrak{C}_1$ is defined by

$$\|\varphi\|_{\mathfrak{C}_1} = \inf \int_{\Omega} \|a_1(\cdot, \omega)\|_{\infty} \|a_2(\cdot, \omega)\|_{\infty} \, d|\nu|(\omega),$$

where the infimum is taken over all possible representations (3.3.3) with $\{a_j(\cdot, \omega)\}_{\omega \in \Omega_k}$ uniformly bounded and uniformly continuous for $j = 1, 2, k \in \mathbb{N}$.

Recall that $\{A_n\}_{n=1}^{\infty} \subset \mathcal{D}_{sa}$ converges to $A \in \mathcal{D}_{sa}$ in the strong resolvent sense if $\{(\lambda I - A_n)^{-1}\}_{n=1}^{\infty}$ converges to $(\lambda I - A)^{-1}$ in the strong operator topology for all λ such that $\mathrm{Im}\, \lambda \neq 0$.

We have the following continuity properties of double operator integrals.

Proposition 3.3.9 *Let $1 \leqslant p \leqslant \infty$. Let $\{A_{i,n}\}_{n=1}^{\infty} \subset \mathcal{D}_{sa}$ converge to $A_i \in \mathcal{D}_{sa}$, $i = 1, 2$, in the strong resolvent sense and let $\{X_n\}_{n=1}^{\infty} \subset \mathcal{S}^p$ ($\mathcal{B}(\mathcal{H})$, respectively) converge to $X \in \mathcal{S}^p$ ($\mathcal{B}(\mathcal{H})$, respectively). The following assertions hold.*

(i) Let $\varphi \in \mathfrak{C}_1$. If $X \in \mathcal{S}^p$, $1 \leqslant p < \infty$, then

$$\lim_{n \to \infty} \|T_{\varphi}^{A_{1,n}, A_{2,n}}(X) - T_{\varphi}^{A_1, A_2}(X)\|_p = 0; \tag{3.3.10}$$

if $X \in \mathcal{B}(\mathcal{H})$, then

$$\text{sot-}\lim_{n\to\infty}(T_\varphi^{A_{1,n},A_{2,n}}(X) - T_\varphi^{A_1,A_2}(X)) = 0. \tag{3.3.11}$$

Moreover, if $\{A_{i,n}\}_{n=1}^\infty$ converge to A_i, $i = 1, 2$, in the operator norm, then the strong operator topology convergence in (3.3.11) can be replaced with the operator norm convergence.

(ii) For every $\varphi \in \mathfrak{A}_1$,

$$\lim_{n\to\infty}\|T_\varphi^{A_1,A_2}(X_n) - T_\varphi^{A_1,A_2}(X)\|_p = 0.$$

(iii) Let $\{\varphi_n\}_{n=1}^\infty \subset \mathfrak{A}_1$ converge to φ in the norm $\|\cdot\|_{\mathfrak{A}_1}$. Then,

$$\lim_{n\to\infty}\|T_{\varphi_n}^{A_1,A_2}(X) - T_\varphi^{A_1,A_2}(X)\|_p = 0.$$

Proof (i) The functions $a_j(\cdot, \omega)$ from the decomposition (3.3.3) are continuous and bounded (uniformly in ω on every Ω_k), so the sequence $\{a_i(A_{n,i}, \omega)\}_n$ converges to $a_i(A_i, \omega)$ in the strong operator topology for every $\omega \in \Omega$, $i = 1, 2$ [171, Theorem VIII.20 (b)]. By [85, Theorem 6.3],

$$\mathcal{S}^p\text{-}\lim_{n\to\infty} a_1(A_{n,1}, \omega)Xa_2(A_{n,2}, \omega) = a_1(A_0, \omega)Xa_2(A_2, \omega),$$

for every $\omega \in \Omega$ if $1 \leqslant p < \infty$. We also have

$$\sup_n \|a_1(A_{n,1}, \cdot)Xa_2(A_{n,2}, \cdot)\|_p \in L^1(\Omega, \mu).$$

Therefore, by the Lebesgue dominated convergence theorem for Bochner integrals, we have the convergence of double operator integrals in the Schatten norm $\|\cdot\|_p$, $1 \leqslant p < \infty$. A completely analogous argument works for convergence in the strong operator topology and operator norm.

The continuity in (ii) and (iii) follows immediately from the estimate (3.3.6). □

The result of Proposition 3.3.9(i) holds for a broader set of symbols φ when $p = 2$. The following fact is established in [55, Proposition 3.1].

Proposition 3.3.10 *Let $\{A_{i,m}\}_{m=1}^\infty \subset \mathcal{D}_{sa}$ converge to $A_i \in \mathcal{D}_{sa}$, $i = 1, 2$, in the strong resolvent sense and let $X \in \mathcal{S}^2$. Then, (3.3.10) with $p = 2$ holds for every $\varphi \in C_b(\mathbb{R}^2)$.*

3.3.6 Symbols of Bounded Double Operator Integrals

There have been many attempts to find an appropriate class of symbols for which the operator T_φ defined in Sect. 3.3.2 is bounded on $\mathcal{B}(\mathcal{H})$. As noted in Sect. 3.3.5,

T_φ is bounded if $\varphi \in \mathfrak{A}_1$. It is straightforward to see that all functions in $\mathfrak{A}_1$ are bounded; therefore, the class $\mathfrak{A}_1$ does not contain polynomials. Further results on boundedness of T_φ are stated below.

Let $1 \leqslant p \leqslant \infty$ and define

$$\mathfrak{M}_p = \mathfrak{M}_p(E_A, E_B) := \{\varphi \in L^\infty(\mathbb{R}^2) : T_\varphi^{A,B} \in \mathcal{B}(S^p)\},$$

$$\mathfrak{M}_{\mathcal{B}(\mathcal{H})} = \mathfrak{M}_{\mathcal{B}(\mathcal{H})}(E_A, E_B) := \{\varphi \in L^\infty(\mathbb{R}^2) : T_\varphi^{A,B} \in \mathcal{B}(\mathcal{B}(\mathcal{H}))\}.$$

It follows from Proposition 3.2.2 that

$$\mathfrak{M}_2(E_A, E_B) = L^\infty(\mathbb{R}^2, G),$$

where G is given by (2.5.1) with $E = E_A$, $F = E_B$.

We also have the following characterization of $\mathfrak{M}_{\mathcal{B}(\mathcal{H})}$, $\mathfrak{M}_1$, and $\mathfrak{M}_\infty$, where the first two equalities can be found in [41, Section 4.1] and the third equality is due to [141, Theorem 1].

Theorem 3.3.11 $\mathfrak{M}_{\mathcal{B}(\mathcal{H})} = \mathfrak{M}_1 = \mathfrak{M}_\infty = \mathfrak{A}_1$.

The description of bounded double operator integrals on $\mathcal{B}(\mathcal{H})$ given by Theorem 3.3.11 is analogous to the celebrated Grothendieck's characterization of bounded Schur multipliers on $\mathcal{B}(\ell^2)$ that can be found in [153, Theorem 5.1].

The class $\mathfrak{M}_p$, $1 < p \neq 2 < \infty$, has not been described yet. However, the following inclusions are known to be strict (see, e.g., [41, (5.7)]):

$$\mathfrak{M}_\infty \subsetneq \mathfrak{M}_p \subsetneq \mathfrak{M}_2. \tag{3.3.12}$$

For instance, it is proved in [159] that for the absolute value function $f(t) = |t|$, its divided difference $f^{[1]}$ belongs to $\mathfrak{M}_p$ whenever $1 < p < \infty$. However, it follows from [80] and [99] that $f^{[1]} \notin \mathfrak{A}_1$, so the left inclusion is strict. Below we demonstrate that the right inclusion in (3.3.12) is also strict.

Let $2 < \alpha < p$,

$$M_d := (e^{-2\pi ijk/d})_{j,k=1}^d, \quad X_d := d^{-1/2-1/p-1/\alpha} \cdot (e^{2\pi ijk/d})_{j,k=1}^d.$$

Since $d^{-1/2}(e^{2\pi ijk/d})_{j,k=1}^d$ is a unitary matrix,

$$\|X_d\|_p = d^{-1/p-1/\alpha}(\mathrm{Tr}(I))^{1/p} = d^{-1/\alpha}.$$

Note that the rescaled Schur product $d^{-1/2+1/p+1/\alpha}(M_d \circ X_d)$ is the orthogonal projection onto the unit vector $d^{-1/2}(1, 1, \ldots, 1) \in \ell_d^2$. Thus,

$$\|M_d \circ X_d\|_p^p = d^{p/2-1-p/\alpha} > d^{-1}.$$

Consider

$$X := \oplus_{d=1}^{\infty} X_d, \quad M := \oplus_{d=1}^{\infty} M_d,$$

the elements of the ℓ^{∞}-direct sum of the matrix algebras $\mathcal{B}(\ell_d^2)$, which embeds in $\mathcal{B}(\mathcal{H})$, where $\mathcal{H}$ is a separable infinite-dimensional Hilbert space. Since $\mathrm{Tr}(X) = \sum_{d=1}^{\infty} \mathrm{Tr}(X_d)$, we have

$$\|X\|_p^p = \sum_{d=1}^{\infty} \|X_d\|_p^p = \sum_{d=1}^{\infty} d^{-p/\alpha} < \infty, \quad \|M \circ X\|_p = \infty.$$

Let A, B be self-adjoint operators such that $\sigma(A)$ consists of the eigenvalues $\{\lambda_j\}_{j=1}^{\infty}$ and $\sigma(B)$ consists of the eigenvalues $\{\mu_k\}_{k=1}^{\infty}$ and let $\varphi \in L^{\infty}(\mathbb{R}^2)$ be such that

$$M = (\varphi(\lambda_j, \mu_k))_{j,k=1}^{\infty}.$$

Then,

$$M \circ X = T_{\varphi}^{A,B}(X)$$

is the double operator integral defined in Sect. 3.2.2. We conclude from the argument above that $T_{\varphi}^{A,B} \notin \mathfrak{M}_p$ for $p > 2$. By the duality argument (see, e.g., [41, (5.6)] or Theorem 4.1.9(iii)),

$$\mathfrak{M}_p = \mathfrak{M}_{p'}, \text{ where } 1 < p, p' < \infty, \ 1/p + 1/p' = 1.$$

Hence, $T_{\varphi}^{A,B} \notin \mathfrak{M}_p$ for $1 < p < 2$ as well. Since, $\varphi \in L^{\infty}(\mathbb{R}^2)$, it follows that $T_{\varphi}^{A,B} \in \mathfrak{M}_2$. One can also generalize the ideas demonstrated above to the case of double operator integrals $T_{\varphi}^{A,B}$, where A, B have arbitrary spectra.

Problem 3.3.12 Describe the class $\mathfrak{M}_p$ for $1 < p \neq 2 < \infty$.

Now we address the question for which functions $f : \mathbb{R} \to \mathbb{C}$ the divided difference $f^{[1]}$ belongs to the class $\mathfrak{A}_1$. Recall that $W_1(\mathbb{R})$ is the Wiener class defined in Sect. 2.1.

Proposition 3.3.13 If $f \in W_1(\mathbb{R})$, then $f^{[1]} \in \mathfrak{A}_1$.

Proof Recalling that $\mathcal{F}^{-1}\mathcal{F}f = f$, we have

$$f^{[1]}(\lambda, \mu) = \frac{f(\lambda) - f(\mu)}{\lambda - \mu} = \frac{1}{\sqrt{2\pi}} \frac{1}{\lambda - \mu} \int_{\mathbb{R}} (e^{i\lambda t} - e^{i\mu t}) \mathcal{F}f(t) \, dt$$

$$= \frac{i}{\sqrt{2\pi}} \int_{\mathbb{R}} \left(\int_0^t e^{i(\lambda s + \mu(t-s))} \, ds \right) \mathcal{F}f(t) \, dt.$$

Making the substitution $u = s$, $v = t - s$ in the latter integral, we obtain

$$f^{[1]}(\lambda, \mu) = \frac{i}{\sqrt{2\pi}} \int_{\mathbb{R}^2} e^{i\lambda u} e^{i\mu v} \mathcal{F} f(u + v) \, du \, dv.$$

Taking $\Omega = \mathbb{R}^2$, $dv(u, v) = \frac{i}{\sqrt{2\pi}} \mathcal{F} f(u + v) \, du \, dv$, $a_1((u, v), \lambda) = e^{i\lambda u}$, and $a_2((u, v), \mu) = e^{i\mu v}$, $u, v \in \mathbb{R}$, we obtain the representation (3.3.3) for the function $f^{[1]}$. We also have $f^{[1]} \in \mathfrak{A}_1$ because

$$\int_{\mathbb{R}} |t \mathcal{F} f(t)| \, dt = \|\mathcal{F} f'\|_{L^1} < \infty.$$

A more delicate harmonic analysis leads to the following result obtained in [141, Theorem 2]:

$$f \in B^1_{\infty 1}(\mathbb{R}) \Rightarrow f^{[1]} \in \mathfrak{A}_1.$$

The above result is sharpened by the theorem below, which is proved in [161, Theorem 5(ii)]. Recall that $\mathfrak{C}_1$ is the class of symbols defined in Sect. 3.3.5, "Continuity".

Theorem 3.3.14 *If $f \in B^1_{\infty 1}(\mathbb{R})$, then $f^{[1]} \in \mathfrak{C}_1$ and*

$$\|f^{[1]}\|_{\mathfrak{C}_1} \leqslant \text{const} \, \|f\|_{B^1_{\infty 1}(\mathbb{R})}.$$

Let $1 \leqslant p \leqslant \infty$ and define

$$\mathscr{F}_p := \{f : \mathbb{R} \to \mathbb{C} : f^{[1]} \in \mathfrak{M}_p\}.$$

In the following theorem $B^1_{11}(\mathbb{R})_{loc}$ denotes the space of functions that belong to the Besov class $B^1_{11}(\mathbb{R})$ locally.

Theorem 3.3.15

(i) $B^1_{11}(\mathbb{R})_{loc} \supsetneq \mathscr{F}_1 = \mathscr{F}_\infty \supsetneq B^1_{\infty 1}(\mathbb{R})$;
(ii) $\mathscr{F}_p = \text{Lip}(\mathbb{R})$, $1 < p < \infty$.

Proof The assertion *(i)* is proved in [141]. Proposition 3.2.2 implies *(ii)* for $p = 2$. The assertion *(ii)* for an arbitrary $1 < p < \infty$ is proved in [159] (see Theorem 5.1.8). □

The result of Theorem 3.3.15(i) was improved in [19], where the authors introduced a new class of functions $\mathfrak{D}$ such that

$$B^1_{\infty 1} \subset \mathfrak{D} \subset \mathscr{F}_1.$$

The fact that the left inclusion above is strict was proved later and can be found in [9, Remark at p. 90]. The original approach of [19] was for the Besov classes on the unit circle. Its analogy for the real line $\mathbb{R}$ can be found in [9, Subsection 3.13].

The class $\mathscr{F}_1$ coincides with the class of operator Lipschitz functions (see [9]). However, the structural properties of the class $\mathscr{F}_1$ in pure scalar terms are not fully understood.

Problem 3.3.16 Describe the class $\mathscr{F}_1$ in terms of familiar function spaces.

3.3.7 Transference Principle

The estimate for the double operator integral $T_{f^{[1]}}$ on $\mathcal{S}^p$, $1 < p < \infty$, given by Theorem 3.3.4 is derived in [159] by transferring boundedness of certain operators on $L^2(\mathbb{R}, X)$ to operators on the UMD space $X = \mathcal{S}^p$, $1 < p < \infty$. The related transference principle and history of transference ideas in general are discussed in [91, Chapter 5]. As noted in Sect. 3.3.6, the double operator integral $T_{|t|^{[1]}}^{A,B}$ is not bounded on $\mathcal{S}^1$ for some choice of A, B with $A - B \in \mathcal{S}^1$. However, as we see below, $T_{|t|^{[1]}}$ is a bounded map from $\mathcal{S}^1$ to $\mathcal{S}^{1,\infty}$ and an estimate similar to the one obtained in Theorem 3.3.4 holds. This estimate for $T_{f^{[1]}}$ in $\mathcal{B}(\mathcal{S}^1, \mathcal{S}^{1,\infty})$ for every f in the largest set of admissible functions $\mathrm{Lip}(\mathbb{R})$ is established in [49, Theorem 1.2]. It follows from deep results in noncommutative analysis and is based on transferring boundedness of singular integral operators on the tensor product space $L^2(\mathbb{T}^2) \otimes \mathcal{H}$ [137] to operators on $\mathcal{H}$.

The following result is due to [49, Theorem 1.2], and it proves the conjecture of [133]. In fact, the result is established in a more general setting for perturbations in the noncommutative L^1-space of a semifinite von Neumann algebra.

Theorem 3.3.17 *Let $A \in \mathcal{D}_{sa}$ and $f \in \mathrm{Lip}(\mathbb{R})$. Then, there exists an absolute constant $c > 0$ such that*

$$\|T_{f^{[1]}}^{A,A}(X)\|_{1,\infty} \leqslant c \, \|f\|_{\mathrm{Lip}(\mathbb{R})} \, \|X\|_1 \tag{3.3.13}$$

for every $X \in \mathcal{S}^1$.

Proof Since the proof of this result is technical, we demonstrate the main idea of the transference method under additional assumptions on A, X, f.

Assume that $A \in \mathcal{B}(\mathcal{H})$ and $\sigma(A)$ is a finite subset of $\mathbb{Z}$, which we denote by $\mathcal{J}$. Let $\{P_j\}_{j \in \mathcal{J}}$ be a sequence of mutually orthogonal projections such that

$$A = \sum_{j \in \mathcal{J}} j P_j, \quad I = \sum_{j \in \mathcal{J}} P_j.$$

Assume that $X \in \mathcal{S}^1$ satisfies $P_j X P_j = 0$ for every $j \in \mathcal{J}$. Finally, assume that $f : \mathbb{Z} \to \mathbb{Z}$.

Define a unitary operator

$$U = \sum_{j \in \mathcal{J}} M_{e_{(j, f(j))}} \otimes P_j \in \mathcal{B}(L^2(\mathbb{T}^2) \otimes \mathcal{H}),$$

where $M_{e_{(m,n)}}$, $m, n \in \mathbb{Z}$, is a multiplication by the function

$$e_{(m,n)}(t_1, t_2) := \exp(-imt_1 - int_2)$$

defined on the torus $\mathbb{T}^2$. Then,

$$
U(I \otimes X)U^* = \left(\sum_{j_3 \in \mathcal{J}} M_{e_{(j_3, f(j_3))}} \otimes P_{j_3} \right) \left(\sum_{j_1, j_2 \in \mathcal{J}} I \otimes P_{j_1} X P_{j_2} \right)
$$
$$
\times \left(\sum_{j_4 \in \mathcal{J}} M_{e_{(-j_4, -f(j_4))}} \otimes P_{j_4} \right)
$$
$$
= \sum_{j_1, j_2 \in \mathcal{J}} M_{e_{(j_1 - j_2, f(j_1) - f(j_2))}} \otimes P_{j_1} X P_{j_2}. \tag{3.3.14}
$$

Similarly,

$$
U\left(I \otimes T^{A,A}_{f^{[1]}}(X) \right)U^* = U\left(I \otimes \sum_{j_1, j_2 \in \mathcal{J}} \frac{f(j_1) - f(j_2)}{j_1 - j_2} P_{j_1} X P_{j_2} \right)U^*
$$
$$
= \sum_{j_1, j_2 \in \mathcal{J}} \frac{f(j_1) - f(j_2)}{j_1 - j_2} M_{e_{(j_1 - j_2, f(j_1) - f(j_2))}} \otimes P_{j_1} X P_{j_2}.
$$
$$\tag{3.3.15}$$

Let $g \in C^1(\mathbb{C} \setminus \{0\})$ be a homogeneous function such that

$$g(t_1 + it_2) = \frac{t_2}{t_1}, \quad |t_2| \leqslant |t_1| \neq 0.$$

Such function g exists and is defined in the proof of [49, Theorem 4.4]. Then, $g(\nabla_{\mathbb{T}^2})$ acts as the Fourier multiplier

$$
g(\nabla_{\mathbb{T}^2})e_{(m-n, f(m)-f(n))} = \frac{f(m) - f(n)}{m - n} e_{(m-n, f(m)-f(n))}, \quad m \neq n \in \mathbb{Z}.
$$
$$\tag{3.3.16}$$

Combining (3.3.14)–(3.3.16) implies

$$U\left(I \otimes T^{A,A}_{f^{[1]}}(X) \right)U^* = (g(\nabla_{\mathbb{T}^2}) \otimes I)(U(I \otimes X)U^*).$$

Hence,

$$\|T^{A,A}_{f^{[1]}}(X)\|_{1,\infty} \leqslant \|g(\nabla_{\mathbb{T}^2}) \otimes I : \mathcal{S}^1 \to \mathcal{S}^{1,\infty}\| \, \|X\|_1.$$

By a particular result of [137] stated in [49, Theorem 2.1] and by [49, Lemma 4.3],

$$g(\nabla_{\mathbb{T}^2}) \otimes I \in \mathcal{B}(\mathcal{S}^1, \mathcal{S}^{1,\infty}),$$

implying (3.3.13) under the additional assumptions on A, X, f.

For the complete proof of (3.3.13) in the general case we refer the reader to [49]. $\square$

3.4 Nonself-adjoint Case

By analogy with double operator integrals for self-adjoint operators defined in Sects. 3.3.3 and 3.3.4, one can define double operator integrals for unitary A and B. The following properties of the transformations discussed in Sect. 3.3.5 extend to the unitary case: the algebraic properties, estimate (3.3.6), perturbation formulas (3.3.7) and (3.3.8), continuity. A unitary analog of the estimate obtained in Theorem 3.3.4 is stated in Theorem 4.3.19 in the general higher order case.

Double operator integrals were introduced for contractions A, B in [142, 147] on the space $\mathcal{S}^2$ similarly to (3.2.1) and on $\mathcal{B}(\mathcal{H})$ similarly to Definition 3.3.2. In the definition of a double operator integral for contractions spectral measures of unitary operators are replaced with semi-spectral measures. Analogous extensions of double operator integrals from self-adjoint to maximal dissipative operators were defined in [6]. When a problem for contractions (respectively, maximal dissipative operators) can be reduced to a problem for unitaries (respectively, self-adjoints) by means of dilations (see, e.g., [132]), one can apply double operator integrals for unitaries without involving the double operator integration theory for contractions (see, e.g., [164]).

3.5 Double Operator Integrals on Noncommutative L^p-Spaces

Double operator integrals were introduced for perturbations in a semi-finite von Neumann algebra $\mathcal{M}$ equipped with a normal semi-finite trace τ. The three constructions analogous to those in the setting of $\mathcal{B}(\mathcal{H})$ are due to [24, 64, 159].

3.5.1 Extension from the Double Operator Integral on $L^2(\mathcal{M}, \tau)$

The double operator integral T_φ^G on the Hilbert space $L^2(\mathcal{M}, \tau)$ can be defined by adjusting Definition 3.2.1. If T_φ^G extends from a dense subset $L^2(\mathcal{M}, \tau) \cap L^p(\mathcal{M}, \tau)$ of $L^p(\mathcal{M}, \tau)$, $1 \leqslant p < \infty$, to a bounded transformation on $L^p(\mathcal{M}, \tau)$, then this extension is called a double operator integral on $L^p(\mathcal{M}, \tau)$. This approach to double operator integration was implemented in [64].

3.5.2 Approach via Separation of Variables

The double operator integral given by Definition 3.3.2 was extended in [24] to perturbations in $\mathcal{M}$ and, in particular, in a symmetrically normed ideal $\mathcal{I}$ of $\mathcal{M}$ with property (F).

A symmetrically normed ideal $\mathcal{I}$ of a semifinite von Neumann algebra $\mathcal{M}$ is said to have property (F) if for every net $\{A_\alpha\}$ in $\mathcal{I}$ satisfying $\sup_\alpha \|A_\alpha\|_{\mathcal{I}} \leqslant 1$ and converging to some $A \in \mathcal{M}$ in the strong operator topology such that $\{A_\alpha^*\}$ converges to A^* in the strong operator topology, it follows that $A \in \mathcal{I}$ and $\|A\|_{\mathcal{I}} \leqslant 1$. Examples of such ideals include $\mathcal{L}^p(\mathcal{M}, \tau) = L^p(\mathcal{M}, \tau) \cap \mathcal{M}$, $1 \leq p < \infty$, and $\mathcal{L}^{p,\infty}(\mathcal{M}, \tau) = L^{p,\infty}(\mathcal{M}, \tau) \cap \mathcal{M}$, $1 < p < \infty$.

It is proved in [24, Lemma 4.6] that the transformation given by Definition 3.3.2 is a bounded map on $\mathcal{I}$.

3.5.3 Approach Without Separation of Variables

The definition (3.3.5) extends to perturbations X in the noncommutative L^p-space $L^p(\mathcal{M}, \tau)$, $1 \leqslant p < \infty$. A more general definition that does not impose restrictions on the spectra of A and B is given in the multilinear setting in Chap. 4.

It is proved in [163, Lemma 3.5] that the multiple operator integrals given by Definition 3.3.2 and formula (3.3.5) coincide for $\varphi \in \mathfrak{C}_1$ on $\mathcal{M}$. They also coincide with the extension of T_φ^G from $\mathcal{L}^2(\mathcal{M}, \tau) \cap \mathcal{L}^p(\mathcal{M}, \tau)$ to $\mathcal{L}^p(\mathcal{M}, \tau)$, $1 \leqslant p < \infty$, discussed in Sect. 3.5.1.

3.5.4   *Properties of Double Operator Integrals on* $L^p(\mathcal{M}, \tau)$

Algebraic Properties

Each of the three constructions of a double operator integral considered in this section satisfies the linearity

$$T_{\alpha_1\varphi_1+\alpha_2\varphi_2} = \alpha_1 T_{\varphi_1} + \alpha_2 T_{\varphi_2}.$$

Further algebraic properties are discussed in Sect. 4.4.3 for general order multiple operator integrals on $L^p(\mathcal{M}, \tau)$.

Norm Estimates

The following bound is a consequence of [24, Lemma 4.6].

Theorem 3.5.1 *Let* $A, B\eta\mathcal{M}_{sa}, \varphi \in \mathfrak{A}_1$, *and let* $\mathcal{I}$ *be a symmetrically normed ideal of* $\mathcal{M}$ *with property* (F). *Then, the double operator integral given by Definition 3.3.2 satisfies*

$$\left\| T_\varphi^{A_1, A_2}(X) \right\|_{\mathcal{I}} \leqslant \|\varphi\|_{\mathfrak{A}_1} \|X\|_{\mathcal{I}}$$

for all $X \in \mathcal{I}$.

The following result is established in [159, Theorem 7].

Theorem 3.5.2 *Let* $A, B\eta\mathcal{M}_{sa}, f \in \mathrm{Lip}(\mathbb{R})$, *and* $1 < p < \infty$. *Then, there exists* $c_p > 0$ *such that*

$$\|T_{f^{[1]}}^{A,B}(X)\|_p \leqslant c_p \|f\|_{\mathrm{Lip}(\mathbb{R})} \|X\|_p$$

for every $X \in L^p(\mathcal{M}, \tau)$.

The following bound is proved in [49, Theorem 1.2].

Theorem 3.5.3 *Let* $A\eta\mathcal{M}_{sa}$ *and* $f \in \mathrm{Lip}(\mathbb{R})$. *Then, there exists an absolute constant* $c > 0$ *such that*

$$\|T_{f^{[1]}}^{A,A}(X)\|_{1,\infty} \leqslant c \|f\|_{\mathrm{Lip}(\mathbb{R})} \|X\|_1$$

for every $X \in L^1(\mathcal{M}, \tau) \cap L^2(\mathcal{M}, \tau)$.

Perturbation Formula

The following analog of Theorem 3.3.8 is obtained in [24, Theorem 5.3].

Theorem 3.5.4 *Let* $A, B\eta M_{sa}$ *be such that* $A - B \in M$ *and let* $f \in W_1(\mathbb{R})$. *Then,*

$$f(A) - f(B) = T^{A,B}_{f^{[1]}}(A - B). \tag{3.5.1}$$

By the method discussed in Theorem 4.4.8 below, one can extend (3.5.1) to $A, B \in (L^q + L^r)(M, \tau)$, $1 < r < q < \infty$, and $f, f' \in C_b(\mathbb{R})$.

3.6 Double Operator Integrals on Banach Spaces

One more interesting direction of development of double operator integration theory is to consider a double operator integral on the space $\mathcal{B}(X, \mathcal{Y})$, where X and $\mathcal{Y}$ are Banach spaces. Such an attempt was firstly made in [64], where the theory of double operator integration was extended in various directions, including the Banach space setting. However, the results in the general setting were much weaker than in the Hilbert space setting. In this section we present results of [173] for scalar type operators on Banach spaces that match analogous results on Hilbert spaces.

Fix Banach spaces X and $\mathcal{Y}$, scalar type operators $A \in \mathcal{B}_s(X)$ and $B \subset \mathcal{B}_s(\mathcal{Y})$ with spectral measures E and F, respectively, and let $\varphi \in \mathfrak{A}_1$. Let a representation as in (3.3.3) for φ be given, with corresponding (Ω, ν) and a_1, a_2. For $\omega \in \Omega$, let $a_1(A, \omega) := a_1(\cdot, \omega)(A) \in \mathcal{B}(X)$ and $a_2(B, \omega) := a_2(\cdot, \omega)(B) \in \mathcal{B}(\mathcal{Y})$ be defined by the functional calculus for A and B, respectively. The next property is due to [173, Lemma 4.1].

Lemma 3.6.1 *Let* $X \in \mathcal{B}(X, \mathcal{Y})$ *have the separable range. Then, for each* $x \in X$, $\omega \mapsto a_2(B, \omega)Xa_1(A, \omega)x$ *is a weakly measurable map* $\Omega \to \mathcal{Y}$.

Now suppose that $\mathcal{Y}$ is separable, I is a Banach ideal in $\mathcal{B}(X, \mathcal{Y})$ and let $X \in \mathcal{B}(X, \mathcal{Y})$. By (2.10.2),

$$\|a_2(B, \omega)Xa_1(A, \omega)\|_I \leqslant 16 \operatorname{spec}(A) \operatorname{spec}(B)\|X\|_I\|a_1(\cdot, \omega)\|_\infty\|a_2(\cdot, \omega)\|_\infty \tag{3.6.1}$$

for $\omega \in \Omega$. Since I is continuously embedded in $\mathcal{B}(X, \mathcal{Y})$, by the Pettis measurability theorem, Lemma 3.6.1, and (3.6.1) we can define the *double operator integral*

$$T^{A,B}_\varphi(X)x := \int_\Omega a_2(B, \omega)Xa_1(A, \omega)x \, d\nu(\omega) \in \mathcal{Y}, \quad x \in X. \tag{3.6.2}$$

The next property is due to [173, Proposition 4.2].

Proposition 3.6.2 *Let X and Y be separable Banach spaces such that X or Y has the bounded approximation property, $A \in \mathcal{B}_s(X)$, $B \in \mathcal{B}_s(Y)$, and $\varphi \in \mathfrak{A}_1$. Let I be a Banach ideal in $\mathcal{B}(X, Y)$ with the strong convex compactness property. Then (3.6.2) defines an operator $T_\varphi^{A,B} \in \mathcal{B}(I)$ which is independent of the choice of representation of φ in (3.3.3), with*

$$\|T_\varphi^{A,B}\|_{I \to I} \leqslant 16 \operatorname{spec}(A) \operatorname{spec}(B) \|\varphi\|_{\mathfrak{A}_1}. \tag{3.6.3}$$

If A and B are self-adjoint operators on separable Hilbert spaces X and Y, then (3.6.3) improves to

$$\|T_\varphi^{A,B}\|_{I \to I} \leqslant \|\varphi\|_{\mathfrak{A}_1}.$$

If $X = Y = \mathcal{H}$ is an infinite-dimensional separable Hilbert space and $I = S^2(\mathcal{H})$, then the definition of a double operator integral on S^2 given in Sect. 3.2 coincides with the definition above for all $\varphi \in \mathfrak{A}_1$ and self-adjoint operators $A, B \in \mathcal{B}(\mathcal{H})$.

Remark 3.6.3 Proposition 3.6.2 cannot in general be extended to a class of functions larger than $\mathfrak{A}_1$. Indeed, as we saw in Theorem 3.3.11, $T_\varphi^{A,B}$ is a bounded operator on $I = \mathcal{B}(\mathcal{H})$ if and only if $\varphi \in \mathfrak{A}_1$. However, for specific Banach ideals, for instance, ideals with the UMD property, results were obtained for larger classes of functions [64, 159] (see Theorem 3.3.15(ii)).

Let us present the following fundamental fact, which is proved in [173, Lemma 4.5], extending the result of [158, Lemma 3] for self-adjoint operators on a separable Hilbert space with difference in a Schatten class. Let $p_1, p_2 : \mathbb{R}^2 \to \mathbb{R}$ be the coordinate projections given by $p_1(\lambda, \mu) := \lambda$, $p_2(\lambda, \mu) := \mu$ for $(\lambda, \mu) \in \mathbb{R}^2$. Note that $f \circ p_1, f \circ p_2 \in \mathfrak{A}_1$ for all bounded Borel functions f.

Lemma 3.6.4 *Under the assumptions of Proposition 3.6.2, the following properties hold:*

(i) *The map $\varphi \mapsto T_\varphi^{A,B}$ is a morphism $\mathfrak{A}_1 \to \mathcal{B}(I)$ of unital Banach algebras.*
(ii) *Let f be a bounded Borel function and $X \in \mathcal{B}(X, Y)$. Then,*

$$T_{f \circ p_1}^{A,B}(X) = Xf(A) \quad and \quad T_{f \circ p_2}^{A,B}(X) = f(B)X.$$

In particular, $T_{p_1}^{A,B}(X) = XA$ and $T_{p_2}^{A,B}(X) = BX$.

This approach to double operator integration theory in the setting of Banach spaces allows to solve a series of interesting problems concerning Lipschitz type estimates in different spaces of operators (see Sect. 5.1.5).

Chapter 4
Multiple Operator Integrals

Theory of multiple operator integrals arose as an extension of the double operator integration theory to the settings that could not be encompassed by the latter constructions. In particular, multilinear transformations naturally arise in finding summable approximations to operator functions in the case of nontrace class perturbations, as we will see in the next chapter. The first attempts to construct suitable multilinear extensions of double operator integrals were made in [138, 196, 199]; the more recent approaches important for applications are due to [24, 56, 146, 163]. In this chapter we discuss the main constructions and properties of multiple operator integrals suitable for applications.

4.1 Multiple Operator Integrals on Finite Matrices

4.1.1 Definition

Let $A_0, \ldots, A_n \in \mathcal{B}_{sa}(\ell_d^2)$, let $\mathfrak{g}^{(j)} = \{\mathfrak{g}_i^{(j)}\}_{i=1}^d$ be an orthonormal basis of eigenvectors of A_j and let $\{\lambda_i^{(j)}\}_{i=1}^d$ be the corresponding d-tuple of eigenvalues, $j = 0, \ldots, n$.

Let $\varphi : \mathbb{R}^{n+1} \to \mathbb{C}$. We define $T_\varphi^{A_0, \ldots, A_n} : \underbrace{\mathcal{B}(\ell_d^2) \times \cdots \times \mathcal{B}(\ell_d^2)}_{n \text{ times}} \to \mathcal{B}(\ell_d^2)$ by

$$T_\varphi^{A_0, \ldots, A_n}(X_1, \ldots, X_n) = \sum_{r_0, \ldots, r_n = 1}^d \varphi\big(\lambda_{r_0}^{(0)}, \lambda_{r_1}^{(1)} \ldots, \lambda_{r_n}^{(n)}\big) P_{\mathfrak{g}_{r_0}^{(0)}} X_1 P_{\mathfrak{g}_{r_1}^{(1)}} \ldots X_n P_{\mathfrak{g}_{r_n}^{(n)}},$$

$$(4.1.1)$$

© Springer Nature Switzerland AG 2019

A. Skripka, A. Tomskova, *Multilinear Operator Integrals*,
Lecture Notes in Mathematics 2250, https://doi.org/10.1007/978-3-030-32406-3_4

for $X_1, \ldots, X_n \in \mathcal{B}(\ell_d^2)$. The operator $T_\varphi^{A_0,\ldots,A_n}$ is a discrete version of a *multiple operator integral*. The function φ is usually called *the symbol* of the operator $T_\varphi^{A_0,\ldots,A_n}$.

Assume that $\{\lambda_i^{(j)}\}_{i=1}^{d_j}$ is the set of pairwise distinct eigenvalues of the operator A_j, where $d_j \in \mathbb{N}$, $d_j \leqslant d$. Then, (4.1.1) can be rewritten in terms of the spectral measures of A_j as

$$T_\varphi^{A_0,\ldots,A_n}(X_1, \ldots, X_n) \tag{4.1.2}$$

$$= \sum_{r_0=1}^{d_0} \cdots \sum_{r_n=1}^{d_n} \varphi(\lambda_{r_0}^{(0)}, \lambda_{r_1}^{(1)} \ldots, \lambda_{r_n}^{(n)}) E_{A_0}(\{\lambda_{r_0}^{(0)}\}) X_1 E_{A_1}(\{\lambda_{r_1}^{(1)}\}) \ldots X_n E_{A_n}(\{\lambda_{r_n}^{(n)}\}).$$

4.1.2 Relation to Multilinear Schur Multipliers

Let $n, d \in \mathbb{N}$ and let

$$\mathfrak{m}(n) := \{m_{r_0,\ldots,r_n}\}_{r_0,\ldots,r_n=1}^{d} \subset \mathbb{C}.$$

An *n-linear Schur multiplier* (or a linear Schur multiplier in case $n = 1$)

$$M_n = M_{\mathfrak{m}(n)} : \underbrace{\mathcal{B}(\ell_d^2) \times \cdots \times \mathcal{B}(\ell_d^2)}_{n \text{ times}} \to \mathcal{B}(\ell_d^2)$$

associated with symbol $\mathfrak{m}(n)$ is defined via

$$M_n(X_1, \ldots, X_n) := \sum_{r_0,\ldots,r_n=1}^{d} m_{r_0,\ldots,r_n} \cdot x_{r_0 r_1}^{(1)} x_{r_1 r_2}^{(2)} \ldots x_{r_{n-1} r_n}^{(n)} \cdot E_{r_0 r_n}, \tag{4.1.3}$$

where

$$X_k = \left(x_{ij}^{(k)}\right)_{i,j=1}^{d}$$

is a $d \times d$ matrix, $1 \leqslant k \leqslant n$. The latter can be rewritten in the form

$$M_n(X_1, \ldots, X_n) = \sum_{r_0,\ldots,r_n=1}^{d} m_{r_0,\ldots,r_n} \cdot E_{r_0} X_1 E_{r_1} X_2 E_{r_2} \ldots X_n E_{r_n}. \tag{4.1.4}$$

Multilinear Schur multipliers whose symbols are divided differences were considered in [73] in the context of perturbation theory. General multilinear Schur multipliers were introduced in [78].

The norm of a multilinear Schur multiplier is estimated via the norm of a linear Schur multiplier in [169, Theorem 2.3].

Theorem 4.1.1 *Let* $1 \leqslant p_1, \ldots, p_n, p \leqslant \infty$ *be such that* $\frac{1}{p_1} + \cdots + \frac{1}{p_n} = \frac{1}{p}$. *For* M_n *defined in (4.1.3) or (4.1.4) and* $M_{1,\tilde{k}}$ *given by*

$$M_{1,\tilde{k}}(C) = \sum_{r,s=1}^{d} m_{r,\underbrace{k,\ldots,k}_{n-1},s} \cdot E_r C E_s,$$

we have

$$\|M_n : S_d^{p_1} \times \cdots \times S_d^{p_n} \to S_d^p\| \geqslant \max_{1 \leqslant k \leqslant d} \|M_{1,\tilde{k}} : S_d^1 \to S_d^p\|.$$

The following result relates multilinear Schur multipliers to multiple operator integrals.

Proposition 4.1.2 *Let* $n \in \mathbb{N}$ *and* $\varphi : \mathbb{R}^{n+1} \to \mathbb{C}$. *Let* $A_0, \ldots, A_n$ *be self-adjoint (or unitary) elements in* $\mathcal{B}(\ell_d^2)$ *and* $X_1, \ldots, X_n \in \mathcal{B}(\ell_d^2)$. *Let* $\mathfrak{g}^{(k)}$ *be an orthonormal basis of eigenvectors of* A_k *with the corresponding d-tuple of eigenvalues* $\{\lambda_l^{(k)}\}_{l=1}^{d}$ *and suppose that* X_k *has the matrix representation* $(x_{ij}^{(k)})_{i,j=1}^{d}$ *in the bases* $\{\mathfrak{g}^{(k)}, \mathfrak{g}^{(k-1)}\}$, $k = 1, \ldots, n$. *The matrix representation of the multiple operator integral with symbol* φ *given by (4.1.1) in the bases* $\{\mathfrak{g}^{(n)}, \mathfrak{g}^{(0)}\}$ *coincides with the value of the Schur multiplier associated with the matrix*

$$\mathfrak{m}(n) = \{\varphi(\lambda_{r_0}^{(0)}, \ldots, \lambda_{r_n}^{(n)})\}_{r_0, \ldots, r_n = 1}^{d}$$

given by (4.1.3) on the n-tuple $\left((x_{ij}^{(1)})_{i,j=1}^{d}, \ldots, (x_{ij}^{(n)})_{i,j=1}^{d}\right)$, *that is,*

$$M_{\mathfrak{m}(n)}(X_1, \ldots, X_n) = T_\varphi^{A_0, \ldots, A_n}(X_1, \ldots, X_n).$$

4.1.3 Properties of Finite Dimensional Multiple Operator Integrals

Algebraic Properties

The following properties of the operator integral $T_\varphi^{A_0, \ldots, A_n}$ are direct consequences of (4.1.4) and Proposition 4.1.2 (for brevity, below we use the notation T_φ):

$$T_{\alpha\varphi + \beta\psi} = \alpha T_\varphi + \beta T_\psi, \quad \alpha, \beta \in \mathbb{C},$$

$$T_\varphi = I, \quad \text{provided} \quad \varphi \equiv 1,$$

where I is the identity operator on $\mathcal{B}(\ell_d^2)$.

Norm Estimates

In this subsection we extend the estimates (3.1.6) and (3.1.7) to the higher order case. The first estimate requires a more delicate care while the second one is obtained by a very minor adjustment of the argument.

Proposition 4.1.3 *Let* $n \in \mathbb{N}$ *and let* $A_0, \ldots, A_n \in \mathcal{B}(\ell_d^2)$ *be self-adjoint (or unitary) operators with corresponding d-tuples of the eigenvalues* $\{\lambda_i^{(j)}\}_{i=1}^d$, $j = 0, \ldots, n$. *Then, for* $\varphi : \mathbb{R}^{n+1} \to \mathbb{C}$ *(or* $\varphi : \mathbb{T}^{n+1} \to \mathbb{C}$*),*

$$\|T_\varphi^{A_0,\ldots,A_n} : S_d^2 \times \cdots \times S_d^2 \to S_d^2\| = \max_{1 \leqslant r_0,\ldots,r_n \leqslant d} |\varphi(\lambda_{r_0}^{(0)}, \lambda_{r_1}^{(1)} \ldots, \lambda_{r_n}^{(n)})|.$$

(4.1.5)

Proof Let $\mathfrak{m}(n) := \{\varphi(\lambda_{r_0}^{(0)}, \ldots, \lambda_{r_n}^{(n)})\}_{r_0,\ldots,r_n=1}^d$. In view of Proposition 4.1.2, in order to prove (4.1.5) it suffices to prove that the Schur multiplier defined in (4.1.3) satisfies

$$\|M_n : S_d^2 \times \cdots \times S_d^2 \to S_d^2\| = \max_{1 \leqslant r_0,\ldots,r_n \leqslant d} |m_{r_0,\ldots,r_n}|.$$

(4.1.6)

Alternatively, one could adjust the method demonstrated below and derive (4.1.5) directly from (4.1.1).

Below we include the proof of the equality (4.1.6) due to [169, Lemma 2.1]. For a fixed $(n+1)$-tuple $(r_0, \ldots, r_n)$, by (4.1.3), we have that

$$M_n(E_{r_0r_1}, E_{r_1r_2}, \ldots, E_{r_{n-1}r_n}) = m_{r_0,\ldots,r_n} E_{r_0r_n}$$

and thus,

$$\|M_n : S_d^2 \times \cdots \times S_d^2 \to S_d^2\| \geqslant |m_{r_0,\ldots,r_n}|$$

for all $1 \leqslant r_0, \ldots, r_n \leqslant d$. Hence, trivially we have

$$\|M_n : S_d^2 \times \cdots \times S_d^2 \to S_d^2\| \geqslant \max_{1 \leqslant r_0,\ldots,r_n \leqslant d} |m_{r_0,\ldots,r_n}|.$$

Conversely, by (4.1.3), we have that

$$\|M_n(X_1, \ldots, X_n)\|_2^2 = \sum_{r_0,r_n=1}^d \left| \sum_{r_1,\ldots,r_{n-1}=1}^d m_{r_0,\ldots,r_n} \cdot x_{r_0r_1}^{(1)} \ldots x_{r_{n-1}r_n}^{(n)} \right|^2$$

(4.1.7)

$$\leqslant \max_{1 \leqslant r_0,\ldots,r_n \leqslant d} |m_{r_0,\ldots,r_n}|^2 \sum_{r_0,r_n=1}^d \left(\sum_{r_1,\ldots,r_{n-1}=1}^d |x_{r_0r_1}^{(1)} \ldots x_{r_{n-1}r_n}^{(n)}| \right)^2,$$

for all $X_1, \ldots, X_n \in S_d^2$. Using the Cauchy-Schwartz inequality n-times, we obtain that

$$
\sum_{r_0, r_n = 1}^{d} \left(\sum_{r_1, \ldots, r_{n-1} = 1}^{d} |x_{r_0 r_1}^{(1)} \ldots x_{r_{n-1} r_n}^{(n)}| \right)^2
$$

$$
= \sum_{r_0, r_n = 1}^{d} \left(\sum_{r_1 = 1}^{d} \left(|x_{r_0 r_1}^{(1)}| \cdot \sum_{r_2, \ldots, r_{n-1} = 1}^{d} |x_{r_1 r_2}^{(2)} \ldots x_{r_{n-1} r_n}^{(n)}| \right) \right)^2
$$

$$
\leqslant \sum_{r_0, r_n = 1}^{d} \sum_{r_1 = 1}^{d} |x_{r_0 r_1}^{(1)}|^2 \cdot \sum_{r_1 = 1}^{d} \left(\sum_{r_2, \ldots, r_{n-1} = 1}^{d} |x_{r_1 r_2}^{(2)} \ldots x_{r_{n-1} r_n}^{(n)}| \right)^2
$$

$$
= \|X_1\|_2^2 \sum_{r_1, r_n = 1}^{d} \left(\sum_{r_2, \ldots, r_{n-1} = 1}^{d} |x_{r_1 r_2}^{(2)} \ldots x_{r_{n-1} r_n}^{(n)}| \right)^2
$$

$$
\leqslant \ldots \leqslant \|X_1\|_2^2 \ldots \|X_n\|_2^2.
$$

Combining the latter with (4.1.7) proves the converse inequality, which completes the proof of (4.1.6) and, hence, of the proposition.   □

Corollary 4.1.4 *Let $n \in \mathbb{N}$ and let $A_0, \ldots, A_n \in \mathcal{B}(\ell_d^2)$ be self-adjoint (or unitary) operators with corresponding d-tuples of the eigenvalues $\{\lambda_i^{(j)}\}_{i=1}^{d}$. Then, for $\varphi : \mathbb{R}^{n+1} \to \mathbb{C}$ (or $\varphi : \mathbb{T}^{n+1} \to \mathbb{C}$) and $X_1, \ldots, X_n \in \mathcal{B}(\ell_d^2)$,*

$$
\|T_\varphi^{A_0, \ldots, A_n}(X_1, \ldots, X_n)\|
$$

$$
\leqslant d^{n/2} \max_{1 \leqslant r_0, \ldots, r_n \leqslant d} |\varphi(\lambda_{r_0}^{(0)}, \lambda_{r_1}^{(1)} \ldots, \lambda_{r_n}^{(n)})| \, \|X_1\| \ldots \|X_n\|. \tag{4.1.8}
$$

We note that the bound (4.1.8) with d^n instead of $d^{n/2}$ trivially follows from the definition of the multilinear Schur multiplier. In some cases such coarse bound is sufficient; see [88, Theorem 2.3.1] or a remark at the end of Sect. 5.3.1.

Perturbation Formula

In the next proposition we establish a standard perturbation formula for a multiple operator integral with respect to auxiliary self-adjoint matrix parameters.

Proposition 4.1.5 *Let* $2 \leqslant n \in \mathbb{N}$, f *be* n *times differentiable on* $\mathbb{R}$ *(or on* $\mathbb{T}$*), and* $A, B, H_1, \ldots, H_{n-1} \in \mathcal{B}(\ell_d^2)$ *be self-adjoint (or unitary) operators. Then,*

$$T_{f^{[k]}}^{H_1,\ldots,H_{j-1},A,H_j,\ldots,H_k}(X_1,\ldots,X_k) - T_{f^{[k]}}^{H_1,\ldots,H_{j-1},B,H_j,\ldots,H_k}(X_1,\ldots,X_k)$$

(4.1.9)

$$= T_{f^{[k+1]}}^{H_1,\ldots,H_{j-1},A,B,H_j,\ldots,H_k}(X_1,\ldots,X_{j-1},A-B,X_j,\ldots,X_k)$$

for all $X_1,\ldots,X_k \in \mathcal{B}(\ell_d^2)$, $0 \leqslant k \leqslant n-1$, $1 \leqslant j \leqslant k+1$.

Proof Fix $n \geqslant 2$ and $0 \leqslant k \leqslant n-1$. For $1 \leqslant j \leqslant k+1$, denote

$$\psi_j(x_0,\ldots,x_{k+1}) := x_j \, f^{[k+1]}(x_0,\ldots,x_{k+1}),$$

$$\phi_j(x_0,\ldots,x_{k+1}) := f^{[k]}(x_0,\ldots,x_{j-1},x_{j+1},\ldots,x_{k+1}).$$

From the definition of the divided difference, we see that

$$\psi_{j-1} - \psi_j = \phi_j - \phi_{j-1}.$$

For brevity, we set

$$\tilde{H}_{j-1} := (H_1,\ldots,H_{j-1}), \quad \tilde{X}_{j-1} := (X_1,\ldots,X_{j-1}),$$

$$_j\tilde{H}_k := (H_j, H_{j+1},\ldots,H_k), \quad _j\tilde{X}_k := (X_j, X_{j+1},\ldots,X_k).$$

By the definition of the multiple operator integral (4.1.1),

$$\text{RHS of (4.1.9)} = T_{f^{[k+1]}}^{\tilde{H}_{j-1},A,B,\,_j\tilde{H}_k}(\tilde{X}_{j-1}, A-B, \,_j\tilde{X}_k)$$

$$= T_{f^{[k+1]}}^{\tilde{H}_{j-1},A,B,\,_j\tilde{H}_k}(\tilde{X}_{j-1}, A, \,_j\tilde{X}_k) - T_{f^{[k+1]}}^{\tilde{H}_{j-1},A,B,\,_j\tilde{H}_k}(\tilde{X}_{j-1}, B, \,_j\tilde{X}_k)$$

$$= T_{\psi_{j-1}}^{\tilde{H}_{j-1},A,B,\,_j\tilde{H}_k}(\tilde{X}_{j-1}, I, \,_j\tilde{X}_k) - T_{\psi_j}^{\tilde{H}_{j-1},A,B,\,_j\tilde{H}_k}(\tilde{X}_{j-1}, I, \,_j\tilde{X}_k)$$

$$= T_{\psi_{j-1}-\psi_j}^{\tilde{H}_{j-1},A,B,\,_j\tilde{H}_k}(\tilde{X}_{j-1}, I, \,_j\tilde{X}_k)$$

$$= T_{\phi_j-\phi_{j-1}}^{\tilde{H}_{j-1},A,B,\,_j\tilde{H}_k}(\tilde{X}_{j-1}, I, \,_j\tilde{X}_k)$$

$$= T_{\phi_j}^{\tilde{H}_{j-1},A,B,\,_j\tilde{H}_k}(\tilde{X}_{j-1}, I, \,_j\tilde{X}_k) - T_{\phi_{j-1}}^{\tilde{H}_{j-1},A,B,\,_j\tilde{H}_k}(\tilde{X}_{j-1}, I, \,_j\tilde{X}_k)$$

$$= T_{f^{[k]}}^{\tilde{H}_{j-1},A,\,_j\tilde{H}_k}(\tilde{X}_k) - T_{f^{[k]}}^{\tilde{H}_{j-1},B,\,_j\tilde{H}_k}(\tilde{X}_k) = \text{LHS of (4.1.9)},$$

proving the assertion.   □

We note that the formula (4.1.9) is proved in [53, Theorem 15] for matrices in the special case $n = 2$.

Continuity

In the next proposition we establish continuity of a multiple operator integral with respect to self-adjoint matrix parameters.

Proposition 4.1.6 *Let* $k \in \mathbb{N}$, $A_j^{(m)}, A_j \in \mathcal{B}(\ell_d^2)$, $j = 1, \ldots, k + 1$, $m \in \mathbb{N}$, *be self-adjoint (respectively, unitary) operators such that* $A^{(m)} \to A$, *as* $m \to \infty$. *Let* $k \in \mathbb{N}$ *and let* $f \in C^k(\mathbb{R})$ *(respectively,* $f \in C^k(\mathbb{T})$*). Then,*

$$
T_{f^{[k]}}^{A^{(m)}, \ldots, A^{(m)}}(X_1, \ldots, X_k) \to T_{f^{[k]}}^{A, \ldots, A}(X_1, \ldots, X_k), \quad m \to \infty,
$$

for all $X_1, \ldots, X_k \in \mathcal{B}(\ell_d^2)$.

Proof We prove the assertion for self-adjoint operators. The case of unitary operators is similar.

Assume first that $f \in C^{k+1}(\mathbb{R})$. Denote

$$
{}_j\tilde{B}_{k+1} := (B_j, B_{j+1}, \ldots, B_{k+1}).
$$

Applying a telescoping summation and Proposition 4.1.5 gives

$$
\left\| T_{f^{[k]}}^{A_1^{(m)}, \ldots, A_{k+1}^{(m)}}(X_1, \ldots, X_k) - T_{f^{[k]}}^{A_1, \ldots, A_{k+1}}(X_1, \ldots, X_k) \right\|_2
$$

$$
= \left\| \sum_{j=1}^{k+1} T_{f^{[k]}}^{{}_1\tilde{A}_{j-1}, \, {}_j\tilde{A}_{k+1}^{(m)}}(X_1, \ldots, X_k) - T_{f^{[k]}}^{{}_1\tilde{A}_j, \, {}_{j+1}\tilde{A}_{k+1}^{(m)}}(X_1, \ldots, X_k) \right\|_2
$$

$$
= \left\| \sum_{j=1}^{k+1} T_{f^{[k+1]}}^{{}_1\tilde{A}_{j-1}, \, A_j^{(m)}, A_j, \, {}_j\tilde{A}_{k+1}^{(m)}}(X_1, \ldots, X_{j-1}, A_j^{(m)} - A_j, X_j, \ldots X_k) \right\|_2.
$$

Applying the triangle inequality, Proposition 4.1.3, and the inequality (2.2.2) (or (2.2.3) for the unitary case), gives

$$
\left\| T_{f^{[k]}}^{A_1^{(m)}, \ldots, A_{k+1}^{(m)}}(X_1, \ldots, X_k) - T_{f^{[k]}}^{A_1, \ldots, A_{k+1}}(X_1, \ldots, X_k) \right\|_2
$$

$$
\leqslant \| f^{(k+1)} \|_\infty \cdot \sum_{j=1}^{k+1} \| A_j^{(m)} - A_j \|_2 \cdot \| X_1 \|_2 \cdots \| X_k \|_2,
$$

which approaches 0 as $m \to \infty$.

Assume now that $f \in C^k(\mathbb{R})$. Given $\epsilon > 0$ there exists $f_\epsilon \in C^{k+1}(\mathbb{R})$ such that

$$\|f^{(k)} - f_\epsilon^{(k)}\|_\infty \|X_1\|_2 \cdots \|X_n\|_2 < \frac{\epsilon}{3}$$

(for instance, f_ϵ could be taken to be a polynomial). Since

$$T_{f^{[k]}}^{A_1^{(m)},\dots,A_{k+1}^{(m)}} - T_{f^{[k]}}^{A_1,\dots,A_{k+1}} = \left(T_{f^{[k]}}^{A_1^{(m)},\dots,A_{k+1}^{(m)}} - T_{f_\epsilon^{[k]}}^{A_1^{(m)},\dots,A_{k+1}^{(m)}}\right)$$

$$+ \left(T_{f_\epsilon^{[k]}}^{A_1^{(m)},\dots,A_{k+1}^{(m)}} - T_{f_\epsilon^{[k]}}^{A_1,\dots,A_{k+1}}\right)$$

$$+ \left(T_{f_\epsilon^{[k]}}^{A_1,\dots,A_{k+1}} - T_{f^{[k]}}^{A_1,\dots,A_{k+1}}\right),$$

by (4.1.5) we obtain

$$\left\|T_{f^{[k]}}^{A_1^{(m)},\dots,A_{k+1}^{(m)}} - T_{f^{[k]}}^{A_1,\dots,A_{k+1}}\right\|_2 \leqslant \frac{2\epsilon}{3} + \left\|(T_{f_\epsilon^{[k]}}^{A_1^{(m)},\dots,A_{k+1}^{(m)}} - T_{f_\epsilon^{[k]}}^{A_1,\dots,A_{k+1}})(X_1,\dots,X_k)\right\|_2.$$
$$\tag{4.1.10}$$

By Proposition 4.1.6 applied to f_ϵ, there exists $m_0 \in \mathbb{N}$ such that for every $m \geqslant m_0$,

$$\left\|(T_{f_\epsilon^{[k]}}^{A_1^{(m)},\dots,A_{k+1}^{(m)}} - T_{f_\epsilon^{[k]}}^{A_1,\dots,A_{k+1}})(X_1,\dots,X_k)\right\|_2 < \frac{\epsilon}{3}. \tag{4.1.11}$$

Combining (4.1.10) and (4.1.11) completes the proof of the proposition in the general case.   □

Remark 4.1.7 We note that the result of Proposition 4.1.6 can also be established by using the following properties of the spectral measures and eigenvalues: for every $\lambda_j \in \sigma(A)$ and $\epsilon > 0$,

$$\lim_{m \to \infty} E_{A_m}((\lambda_j - \epsilon, \lambda_j + \epsilon)) = E_A((\lambda_j - \epsilon, \lambda_j + \epsilon))$$

and there exist $\lambda_j^{(m)} \in \sigma(A_m)$, $m \in \mathbb{N}$, such that

$$\lim_{m \to \infty} \lambda_j^{(m)} = \lambda_j$$

(see, e.g., [171, Theorem VIII.24]). However, this property extends to the infinite dimensional case only under the restriction that $\lambda_j - \epsilon$, $\lambda_j + \epsilon$ are not eigenvalues of A. On the other hand, the ideas used to prove Proposition 4.1.6 generalize to the infinite dimensional setting.

Reduction to Identical Spectral Measures

Below we prove a useful property of multiple operator integrals, which shows that instead of working with $T_\varphi^{B,A,\dots,A}$ defined in (4.1.1) for $A \neq B$, it suffices to work with $T_\varphi^{A,A,\dots,A}$.

Proposition 4.1.8 *Let $A, B \in \mathcal{B}(\ell_d^2)$ be self-adjoint operators and $X_1, \dots, X_n \in \mathcal{B}(\ell_d^2)$. Let E_{ij} denote the elementary 2×2 matrix whose only nonzero entry equals 1 and has indices (i, j) and denote*

$$C = E_{11} \otimes A + E_{22} \otimes B, \quad \tilde{X}_1 = E_{12} \otimes X_1, \quad and \quad \tilde{X}_j = E_{22} \otimes X_j, \quad j = 2, \dots, n.$$

Then, for a bounded Borel function $\varphi : \mathbb{R}^{n+1} \to \mathbb{C}$,

$$T_\varphi^{C,\dots,C}(\tilde{X}_1, \dots, \tilde{X}_n) = E_{12} \otimes T_\varphi^{A,B,\dots,B}(X_1, \dots, X_n).$$

Proof Let $\sigma(A) = \{\lambda_i\}_{i=1}^m$ and $\sigma(B) = \{\lambda'_j\}_{j=1}^{m'}$ for $m, m' \leqslant d$. The spectral projection of the operator C associated with $\lambda \in \sigma(A) \cup \sigma(B)$ is given by

$$E_C(\lambda) = E_{11} \otimes E_A(\lambda) + E_{22} \otimes E_B(\lambda).$$

By (4.1.2),

$$T_\varphi^{C,\dots,C}(\tilde{X}_1, \dots, \tilde{X}_n) = \sum_{\lambda_{r_0},\dots,\lambda_{r_n} \in \sigma(A) \cup \sigma(B)} \varphi(\lambda_{r_0}, \dots, \lambda_{r_n}) E_C(\lambda_{r_0}) \tilde{X}_1 \dots \tilde{X}_n E_C(\lambda_{r_n}).$$

Moreover,

$$
\begin{aligned}
&E_C(\lambda_{r_0}) \tilde{X}_1 \dots \tilde{X}_n E_C(\lambda_{r_n}) \\
&= \big(E_{11} \otimes E_A(\lambda_{r_0}) + E_{22} \otimes E_B(\lambda_{r_0})\big)(E_{12} \otimes X_1) E_C(\lambda_{r_1}) \tilde{X}_2 \dots \tilde{X}_n E_C(\lambda_{r_n}) \\
&= \big(E_{12} \otimes E_A(\lambda_{r_0}) X_1\big) E_C(\lambda_{r_1})(E_{22} \otimes X_2) E_C(\lambda_{r_2}) \tilde{X}_3 \dots \tilde{X}_n E_C(\lambda_{r_n}) \\
&= \big(E_{12} \otimes E_A(\lambda_{r_0}) X_1 E_B(\lambda_{r_1}) X_2\big) E_C(\lambda_{r_2}) \tilde{X}_3 \dots \tilde{X}_n E_C(\lambda_{r_n}) \\
&= \dots = E_{12} \otimes \big(E_A(\lambda_{r_0}) X_1 E_B(\lambda_{r_1}) X_2 \dots X_n E_B(\lambda_{r_n})\big).
\end{aligned}
$$

Therefore,

$$
\begin{aligned}
&T_\varphi^{C,\dots,C}(\tilde{X}_1, \dots, \tilde{X}_n) \\
&= \sum_{\lambda_{r_0} \in \sigma(A), \lambda_{r_1} \dots, \lambda_{r_n} \in \sigma(B)} \varphi(\lambda_{r_0}, \dots, \lambda_{r_n}) E_{12} \otimes \big(E_A(\lambda_{r_0}) X_1 E_B(\lambda_{r_1}) X_2 \dots X_n E_B(\lambda_{r_n})\big) \\
&= E_{12} \otimes T_\varphi^{A,B,\dots,B}(X_1, \dots, X_n).
\end{aligned}
$$

4.1.4 Estimates of Multiple Operator Integrals via Double Operator Integrals

The theorem below reduces estimates for multiple operator integrals to estimates for double operator integrals.

Theorem 4.1.9 *Let* $1 \leqslant p_1, \ldots, p_n, p \leqslant \infty$ *be such that* $\frac{1}{p_1} + \cdots + \frac{1}{p_n} = \frac{1}{p}$ *and let* $\psi : \mathbb{C}^{n+1} \to \mathbb{C}$ *be a bounded Borel function. Let* $A, B \in \mathcal{B}(\ell_d^2)$. *The following assertions hold.*

(i) *If A and B are self-adjoint,* $0 \in \sigma(A)$, *and*

$$\varphi(x_0, x_n) := \psi(x_0, 0, \ldots, 0, x_n), \quad \text{for } x_0, x_n \in \mathbb{R},$$

then

$$\|T_\psi^{A+B,A,\ldots,A} : \mathcal{S}_d^{p_1} \times \cdots \times \mathcal{S}_d^{p_n} \to \mathcal{S}_d^p\| \geqslant \|T_\varphi^{A+B,A} : \mathcal{S}_d^1 \to \mathcal{S}_d^p\|.$$

(ii) *If A and A + B are unitary,* $1 \in \sigma(A)$, *and*

$$\varphi(z_0, z_n) := \psi(z_0, 1, \ldots, 1, z_n), \quad \text{for } z_0, z_n \in \mathbb{T},$$

then

$$\|T_\psi^{A+B,A,\ldots,A} : \mathcal{S}_d^{p_1} \times \cdots \times \mathcal{S}_d^{p_n} \to \mathcal{S}_d^p\| \geqslant \|T_\varphi^{A+B,A} : \mathcal{S}_d^1 \to \mathcal{S}_d^p\|.$$

(iii) *If* $n = 1$, $1 \leqslant p, p' \leqslant \infty$, *and* $\frac{1}{p} + \frac{1}{p'} = 1$, *then*

$$\|T_\psi^{A+B,A} : \mathcal{S}_d^p \to \mathcal{S}_d^p\| = \|T_\psi^{A+B,A} : \mathcal{S}_d^{p'} \to \mathcal{S}_d^{p'}\|.$$

Proof The properties (i) and (ii) are immediate consequences of Theorem 4.1.1 and Proposition 4.1.2. The property (iii) follows from the representation (4.1.4) and Proposition 4.1.2. □

4.2 Multiple Operator Integrals on $\mathcal{S}^2$

4.2.1 Pavlov's Approach

Let $A_1, \ldots, A_{n+1} \in \mathcal{D}_{sa}$ and $X_1 \ldots, X_n \in \mathcal{S}^2$. Let $\Omega = \sigma(A_1) \times \cdots \times \sigma(A_{n+1})$ and δ_k be a Borel subset of $\sigma(A_k)$, $k = 1, \ldots, n+1$. Consider the finitely additive $\mathcal{S}^2$-valued measure

$$m(\delta_1 \times \cdots \times \delta_{n+1}) = E_{A_1}(\delta_1) X_1 \ldots X_n E_{A_{n+1}}(\delta_{n+1})$$

defined on finite unions of the sets $\delta_1 \times \cdots \times \delta_{n+1}$, which we call rectangular subsets of $\sigma(A_1) \times \cdots \times \sigma(A_{n+1})$. It is proved in [138, Theorem 1] that m has a bounded semivariation

$$\|m\| \leqslant \|X_1\|_2 \ldots \|X_n\|_2$$

and is σ-additive. Since the space $\mathcal{S}^2$ is reflexive, it follows from [66, Chapter 1, Section 5, Theorem 2] that m has a unique σ-additive extension $\tilde{m} : \Sigma \to \mathcal{S}^2$, where Σ is the σ-algebra generated by the rectangular subsets of $\sigma(A_1) \times \cdots \times \sigma(A_{n+1})$. The measure $\tilde{m}$ has a bounded semivariation and is absolutely continuous with respect to the direct product $\lambda_{A_1} \times \cdots \times \lambda_{A_{n+1}}$ of scalar-valued spectral measures of $A_1, \ldots, A_{n+1}$. Hence, $L^\infty(\Omega, \lambda_{A_1} \times \cdots \times \lambda_{A_{n+1}}) \subset L^\infty(\Omega, \tilde{m})$ and, by [66, Chapter 1, Section 1, Theorem 13], the $\mathcal{S}^2$-valued integral

$$\int_\Omega \varphi(\omega) \, d\tilde{m}(\omega) \tag{4.2.1}$$

is well defined for $\varphi \in L^\infty(\Omega, \lambda_{A_1} \times \cdots \times \lambda_{A_{n+1}})$. This integral is called in [138] a multiple operator integral associated with φ, $A_1, \ldots, A_{n+1}$, and $X_1, \ldots, X_n$.

If $n = 1$, then $\tilde{m}(E_{A_1}(\delta_1) X E_{A_2}(\delta_2)) = G(\delta_1 \times \delta_2)$ for every $X \in \mathcal{S}^2$, where G is given by (2.5.1). Thus, the multiple operator integral given by (4.2.1) extends the double operator integral given by (3.2.4).

4.2.2 Coine-Le Merdy-Sukochev's Approach

The following multiple operator integral is defined in [56]. Given a self-adjoint operator $A \in \mathcal{D}_{sa}$, let λ_A denote its scalar-valued spectral measure. For $A_1, \ldots, A_{n+1} \in \mathcal{D}_{sa}$, let $\Omega = \sigma(A_1) \times \cdots \times \sigma(A_{n+1})$.

Definition 4.2.1 Let Γ be the unique linear map from the tensor product space $L^\infty(\sigma(A_1), \lambda_1) \otimes \cdots \otimes L^\infty(\sigma(A_{n+1}), \lambda_{n+1})$ into $\mathcal{B}_n(\mathcal{S}^2 \times \cdots \times \mathcal{S}^2, \mathcal{S}^2)$ such that

$$\Gamma(f_1 \otimes \cdots \otimes f_{n+1})(X_1, \ldots, X_n) = f_1(A_1) X_1 f_2(A_2) X_2 \cdots f_n(A_n) X_n f_{n+1}(A_{n+1})$$

for all $f_j \in L^\infty(\sigma(A_j), \lambda_j)$, $j = 1, \ldots, n$, and all $X_1, \ldots, X_n \in \mathcal{S}^2$. According to [56, Proposition 6], Γ uniquely extends to a w^*-continuous and contractive map

$$\Gamma^{A_1, \ldots, A_{n+1}} : L^\infty(\Omega, \lambda_{A_1} \times \cdots \times \lambda_{A_{n+1}}) \longrightarrow \mathcal{B}_n(\mathcal{S}^2 \times \cdots \times \mathcal{S}^2, \mathcal{S}^2).$$

Let $\varphi : \mathbb{R}^{n+1} \to \mathbb{C}$ be a bounded Borel function and let $\tilde{\varphi} \in L^\infty(\Omega, \lambda_{A_1} \times \cdots \times \lambda_{A_{n+1}})$ be the class of its restriction to Ω. Denote the n-linear map $\Gamma^{A_1, \ldots, A_{n+1}}(\tilde{\varphi})$ by

$$\Gamma^{A_1, \ldots, A_{n+1}}(\varphi) : \mathcal{S}^2 \times \cdots \times \mathcal{S}^2 \longrightarrow \mathcal{S}^2.$$

We note that $\mathcal{B}_n(\mathcal{S}^2 \times \cdots \times \mathcal{S}^2, \mathcal{S}^2)$ is a dual space, and the predual is given by

$$\mathcal{S}^2 \overset{\wedge}{\otimes} \cdots \overset{\wedge}{\otimes} \mathcal{S}^2,$$

the projective tensor product of n copies of $\mathcal{S}^2$ [66]. The w^*-continuity of $\Gamma^{A_1,\dots,A_{n+1}}$ means that if a net $\{\varphi_i\}_{i \in I}$ in $L^\infty \left(\Omega, \prod_{i=1}^{n+1} \lambda_{A_i} \right)$ converges to $\varphi \in L^\infty \left(\Omega, \prod_{i=1}^{n+1} \lambda_{A_i} \right)$ in the w^*-topology, then for all $X_1, \dots, X_n \in \mathcal{S}^2$, the net

$$\left\{ \Gamma^{A_1,\dots,A_{n+1}}(\varphi_i)(X_1, \dots, X_n) \right\}_{i \in I}$$

converges to $\Gamma^{A_1,\dots,A_{n+1}}(\varphi)(X_1, \dots, X_n)$ weakly in $\mathcal{S}^2$.

The following connection between different multiple operator integrals is established in [56, Remark 8].

Proposition 4.2.2 *The multiple operator integral* $\Gamma(\varphi)$ *given by Definition 4.2.1 coincides with Pavlov's multiple operator integral given by* (4.2.1).

The crucial point is the construction leading to Definition 4.2.1 is the w^*-continuity of $\Gamma^{A_1,\dots,A_{n+1}}(\cdot)$, which allows to reduce various computations to elementary tensor product manipulations. These ideas are illustrated in [55].

The symbols φ for which the triple operator integral $\Gamma^{A_1,A_2,A_3}(\varphi)$ maps $\mathcal{S}^2 \times \mathcal{S}^2$ to $\mathcal{S}^1$ are characterized in [56, Theorem 23]. In fact, this characterization is obtained for a natural generalization of $\Gamma^{A_1,A_2,A_3}(\varphi)$ to normal operators A_1, A_2, A_3.

Theorem 4.2.3 *Let* A, B *and* C *be normal operators densely defined in* $\mathcal{H}$ *and let* $\phi \in L^\infty(\sigma(A) \times \sigma(B) \times \sigma(C), \lambda_A \times \lambda_B \times \lambda_C)$. *The following are equivalent:*

(i) $\Gamma^{A,B,C}(\phi) \in \mathcal{B}_2(\mathcal{S}^2 \times \mathcal{S}^2, \mathcal{S}^1)$.
(ii) There exist a separable Hilbert space H *and two functions*

$$a \in L^\infty(\sigma(A) \times \sigma(B), \lambda_A \times \lambda_B; H) \quad and \quad b \in L^\infty(\sigma(B) \times \sigma(C), \lambda_B \times \lambda_C; H)$$

such that

$$\phi(t_1, t_2, t_3) = \langle a(t_1, t_2), b(t_2, t_3) \rangle$$

for a.e. $(t_1, t_2, t_3) \in \sigma(A) \times \sigma(B) \times \sigma(C)$.

In this case,

$$\|\Gamma^{A,B,C}(\phi) : \mathcal{S}^2 \times \mathcal{S}^2 \longrightarrow \mathcal{S}^1\| = \inf\{\|a\|_\infty \|b\|_\infty\},$$

where the infimum runs over all pairs (a, b) *satisfying (ii).*

4.3 Multiple Operator Integrals on Schatten Classes and $\mathcal{B}(\mathcal{H})$

4.3.1 Approach via Separation of Variables

In this subsection we extend the definition of Sect. 3.3.3 to the higher order case. This approach to multiple operator integral was developed in [146], [24] and summarized in [149].

Let $n \in \mathbb{N}$ and let $\mathfrak{A}_n$ be the class of functions $\varphi : \mathbb{R}^{n+1} \to \mathbb{C}$ admitting the representation

$$\varphi(\lambda_1, \ldots, \lambda_{n+1}) = \int_\Omega a_1(\lambda_1, \omega) \ldots a_{n+1}(\lambda_{n+1}, \omega) \, dv(\omega), \qquad (4.3.1)$$

for some finite measure space (Ω, ν) and bounded measurable functions

$$a_i(\cdot, \cdot) : \mathbb{R} \times \Omega \to \mathbb{C}, \quad i = 1, \ldots, n + 1,$$

where on $\mathbb{R}$ we consider the Borel σ-algebra, such that

$$\int_\Omega \|a_1(\cdot, \omega)\|_\infty \ldots \|a_{n+1}(\cdot, \omega)\|_\infty \, d\,|\nu|\,(\omega) < \infty.$$

The class $\mathfrak{A}_n$ is an algebra with respect to the operations of pointwise addition and multiplication [24, Proposition 4.10]. The formula

$$\|\varphi\|_{\mathfrak{A}_n} = \inf \int_\Omega \|a_1(\cdot, \omega)\|_\infty \ldots \|a_{n+1}(\cdot, \omega)\|_\infty \, d\,|\nu|\,(\omega),$$

where the infimum is taken over all possible representations (4.3.1), defines a norm on $\mathfrak{A}_n$ (see [63, Lemma 4.6]).

The class $\mathfrak{A}_n$ can also be defined as the class of functions $\varphi : \mathbb{R}^{n+1} \to \mathbb{C}$ admitting the representation

$$\varphi(\lambda_1, \ldots, \lambda_{n+1}) = \int_\Omega b_1(\lambda_1, \omega) \ldots b_{n+1}(\lambda_{n+1}, \omega) \, dv_2(\omega), \qquad (4.3.2)$$

for some (not necessarily finite) measure space (Ω, ν_2) and bounded measurable functions

$$b_i(\cdot, \cdot) : \mathbb{R} \times \Omega \to \mathbb{C}, \quad i = 1, \ldots, n + 1,$$

such that

$$\int_\Omega \|b_1(\cdot, \omega)\|_\infty \ldots \|b_{n+1}(\cdot, \omega)\|_\infty \, d\,|\nu_2|\,(\omega) < \infty$$

(see, e.g., [63, 158]). These definitions coincide since, as in the case of double operator integrals, the representation (4.3.1) of the function φ can be obtained from (4.3.2) with

$$a_1(\lambda, \omega) = \frac{b_1(\lambda, \omega)}{\|b_1(\cdot, \omega)\|_\infty}, \quad \ldots, \quad a_{n+1}(\lambda, \omega) = \frac{b_{n+1}(\lambda, \omega)}{\|b_{n+1}(\cdot, \omega)\|_\infty}$$

and the finite measure ν defined by

$$\nu = \|b_1(\cdot, \omega)\|_\infty \cdots \|b_{n+1}(\cdot, \omega)\|_\infty \nu_2.$$

The following definition was introduced independently in [24, Definition 4.1] and [146, (3.9)].

Definition 4.3.1 Let $1 \leqslant p, p_1, \ldots, p_n \leqslant \infty$. For every $\varphi \in \mathfrak{A}_n$, and a fixed tuple $A_1, \ldots, A_{n+1} \in \mathcal{D}_{sa}$, the double operator integral $T_\varphi^{A_1, \ldots, A_{n+1}} : S^{p_1} \times \cdots \times S^{p_n} \to S^p$, where $\frac{1}{p_1} + \cdots + \frac{1}{p_n} = \frac{1}{p}$ (respectively, $T_\varphi^{A_1, \ldots, A_{n+1}} : \mathcal{B}(\mathcal{H}) \times \cdots \times \mathcal{B}(\mathcal{H}) \to \mathcal{B}(\mathcal{H})$) is defined by

$$T_\varphi^{A_1, \ldots, A_{n+1}}(X_1, \ldots, X_n) := \int_\Omega a_1(A_1, \omega) X_1 \ldots X_n a_{n+1}(A_{n+1}, \omega) \, d\nu(\omega),$$

for $X_j \in S^{p_j}$ (respectively, $X_j \in \mathcal{B}(\mathcal{H})$ if $p_j = \infty$), where a_j and (Ω, ν) are taken from the representation (4.3.1) and the integral is understood in the sense of the Bochner integral

$$\left(\int_\Omega a_1(A_1, \omega) X_1 \ldots X_n a_{n+1}(A_{n+1}, \omega) \right) d\nu(\omega)(y)$$

$$= \int_\Omega (a_1(A_1, \omega) X_1 \ldots X_n a_{n+1}(A_{n+1}, \omega))(y) \, d\nu(\omega), \quad y \in \mathcal{H}.$$

It is worth noting that the multiple operator integral is introduced above for the symbol φ from the integral projective tensor product of L^∞-spaces which coincides with the extended Haagerup tensor product in the case $n = 2$. The definition and study of the multiple operator integral with respect to the symbol from the extended Haagerup tensor product for $n > 2$ can be found in [95] and [11]; see also [14].

The value $T_\varphi^{A_1, \ldots, A_{n+1}}(X_1, \ldots, X_n)$ does not depend on the particular representation on the right-hand side of (4.3.1). This fact is stated in [146, Lemma 3.1] and proved under the assumption $a_i(\cdot, \omega)$ be bounded and continuous for every $\omega \in \Omega$, $i = 1, \ldots, n + 1$, in [24, Lemma 4.3].

It follows directly from Definition 4.3.1 that if $\varphi \in \mathfrak{A}_n$, then $T_\varphi^{A_1, \ldots, A_{n+1}}$ is a bounded n-linear transformation that maps $S^{p_1} \times \cdots \times S^{p_n} \to S^p$, $1 \leqslant p, p_1, \ldots, p_n \leqslant \infty$, and $\mathcal{B}(\mathcal{H}) \times \cdots \times \mathcal{B}(\mathcal{H}) \to \mathcal{B}(\mathcal{H})$.

The following connection between different multiple operator integrals is established in [56, Proposition 28].

Proposition 4.3.2 *Let $A_1, \ldots, A_{n+1} \in \mathcal{D}_{sa}$ and $\varphi \in \mathfrak{A}_n$. Then, the multiple operator integral $\Gamma^{A_1, \ldots, A_{n+1}}(\varphi)$ given by Definition 4.2.1 coincides on $\mathcal{S}^2 \times \cdots \times \mathcal{S}^2$ with the multiple operator integral $T_\varphi^{A_1, \ldots, A_{n+1}}$ given by Definition 4.3.1.*

4.3.2 Approach Without Separation of Variables

In this section we extend the definition of Sect. 4.1.1 to the case of infinite-dimensional operators. The multiple operator integral that we introduce below is different from the one introduced in Sect. 4.3.1; however, the two constructions coincide for a broad set of symbols used in applications.

Let $A_j \in \mathcal{D}_{sa}$, $j = 0, \ldots, n$, and denote

$$E_{l,m}^j = E_{A_j}\left(\left[\frac{l}{m}, \frac{l+1}{m}\right)\right),$$

for every $m \in \mathbb{N}$, $l \in \mathbb{Z}$, and $j = 0, \ldots, n$. Let $n \in \mathbb{N}$ and $1 \leqslant p, p_j \leqslant \infty$, $j = 1, \ldots, n$, be such that $0 \leqslant \frac{1}{p} = \frac{1}{p_1} + \cdots + \frac{1}{p_n} \leqslant 1$. Let $X_j \in \mathcal{S}^{p_j}$ and let $\varphi : \mathbb{R}^{n+1} \to \mathbb{C}$ be a bounded Borel function. The following definition is due to [163, Definition 3.1].

Definition 4.3.3 Suppose that for every $m \in \mathbb{N}$, the series

$$S_{\varphi,m}(X_1, \ldots, X_n) := \sum_{l_0, \ldots, l_n \in \mathbb{Z}} \varphi\left(\frac{l_0}{m}, \ldots, \frac{l_n}{m}\right) E_{l_0,m}^0 X_1 E_{l_1,m}^1 X_2 \ldots X_n E_{l_n,m}^n$$

$$= \lim_{N \to \infty} \sum_{\substack{|l_j| \leqslant N \\ 0 \leqslant j \leqslant n}} \varphi\left(\frac{l_0}{m}, \ldots, \frac{l_n}{m}\right) E_{l_0,m}^0 X_1 E_{l_1,m}^1 X_2 \ldots X_n E_{l_n,m}^n$$

converges in the norm of $\mathcal{S}^p$ and

$$(X_1, \ldots, X_n) \mapsto S_{\varphi,m}(X_1, \ldots, X_n), \quad m \in \mathbb{N},$$

is a sequence of bounded multilinear operators that map $\mathcal{S}^{p_1} \times \cdots \times \mathcal{S}^{p_n} \to \mathcal{S}^p$. If the sequence of operators $\{S_{\varphi,m}\}_{m=1}^\infty$ converges strongly to some multilinear operator T_φ, then, according to the Banach-Steinhaus theorem, $\{S_{\varphi,m}\}_{m=1}^\infty$ is uniformly bounded and the operator T_φ is also bounded. The transformation T_φ is called the multiple operator integral associated with φ and the operators $A_0, \ldots, A_n$ (or the spectral measures $E_{A_0}, \ldots, E_{A_n}$).

Let $\mathfrak{C}_n$ denote the subset of $\mathfrak{A}_n$ of functions admitting the representation (4.3.1), where $\cup_{k=1}^\infty \Omega_k = \Omega$ for a growing sequence $\{\Omega_k\}_{k=1}^\infty$ of measurable subsets of Ω such that the families $\{a_j(\cdot, \omega)\}_{\omega \in \Omega_k}$, $j = 1, \ldots, n+1$, are uniformly bounded and uniformly continuous. A norm on $\mathfrak{C}_n$ is defined by

$$\|\varphi\|_{\mathfrak{C}_n} = \inf \int_\Omega \|a_1(\cdot, \omega)\|_\infty \cdots \|a_{n+1}(\cdot, \omega)\|_\infty \, d\,|\nu|\,(\omega),$$

where the infimum is taken over all possible representations (4.3.1) with a_j, $j = 1, \ldots, n+1$, as above. Multiple operator integrals with symbols in $\mathfrak{C}_n$ are known to enjoy nicer properties than multiple operator integrals with symbols in $\mathfrak{A}_n$.

The following sufficient condition for $\varphi \in \mathfrak{C}_n$ is established in [161, Theorem 5]. A more general set of symbols in $\mathfrak{C}_n$ will be considered in Sect. 4.3.3.

Theorem 4.3.4 *If* $f \in B_{\infty 1}^n(\mathbb{R})$, *then* $f^{[n]} \in \mathfrak{C}_n$ *and*

$$\|f^{[n]}\|_{\mathfrak{C}_n} \leqslant \mathrm{const}_n \, \|f\|_{B_{\infty 1}^n(\mathbb{R})}.$$

The following result is established in [163, Lemma 3.5].

Proposition 4.3.5 *The multiple operator integrals* T_φ *given by Definitions 4.3.1 and 4.3.3 coincide for* $\varphi \in \mathfrak{C}_n$.

The multiple operator integral on a Cartesian product of Hilbert-Schmidt ideals defined in this subsection coincides with the one given by Definition 4.2.1, as obtained in [119, Lemma 2.6 and Remark 2.7(i)].

Proposition 4.3.6 *The multiple operator integrals* T_φ *and* $\Gamma(\varphi)$ *given by Definitions 4.3.3 and 4.2.1 coincide on* $\mathcal{S}^2 \times \cdots \times \mathcal{S}^2$ *for every* $\varphi \in C_b(\mathbb{R}^{n+1})$.

Proof Let $n \in \mathbb{N}$, let $A_1, \ldots, A_{n+1}$ be self-adjoint operators, $X_1, \ldots, X_n \in \mathcal{S}^2$, and $\varphi \in C_b(\mathbb{R}^{n+1})$. It can be shown that $T_\varphi^{A_1, \ldots, A_{n+1}} : \mathcal{S}^2 \times \cdots \times \mathcal{S}^2 \to \mathcal{S}^2$ is well defined.

For any $r \in \mathbb{N}, l \in \mathbb{Z}$, set $J_{l,r} = \left[\frac{l}{r}, \frac{l+1}{r}\right)$ and for $N \in \mathbb{N}$, consider

$$\varphi_{r,N} = \sum_{\substack{|l_j| \leqslant N \\ 1 \leqslant j \leqslant n+1}} \varphi_{r,N}\left(\frac{l_1}{r}, \ldots, \frac{l_{n+1}}{r}\right) \chi_{J_{l_1,r}} \otimes \cdots \otimes \chi_{J_{l_{n+1},r}}.$$

Since $\varphi \colon \mathbb{R}^{n+1} \to \mathbb{C}$ is continuous,

$$\varphi = w^*\text{-}\lim_{r\to\infty} \lim_{N\to\infty} \varphi_{r,N}$$

in $L^\infty(\Omega, \lambda_{A_1} \times \cdots \times \lambda_{A_{n+1}})$. Hence for all $X_1, \ldots, X_n \in \mathcal{S}^2$,

$$\Gamma^{A_1, \ldots, A_{n+1}}(\varphi)(X_1, \ldots, X_n) = \lim_{r\to\infty} \lim_{N\to\infty} \Gamma^{A_1, \ldots, A_{n+1}}(\varphi_{r,N})(X_1, \ldots, X_n)$$

in $\mathcal{S}^2$. Comparing the latter with Definition 4.3.3 implies

$$\Gamma^{A_1,\ldots,A_{n+1}}(\varphi)(X_1,\ldots,X_n) = T_\varphi^{A_1,\ldots,A_{n+1}}(X_1,\ldots,X_n). \qquad (4.3.3)$$

4.3.3   Properties of Multiple Operator Integrals on $\mathcal{S}^p$ and $\mathcal{B}(\mathcal{H})$

Algebraic Properties

The multiple operator integral given by Definition 4.3.1 satisfies

$$T_{\alpha_1\varphi_1+\alpha_2\varphi_2} = \alpha_1 T_{\varphi_1} + \alpha_2 T_{\varphi_2}, \qquad (4.3.4)$$

for $\varphi_1, \varphi_2 \in \mathfrak{A}_n, \alpha_1, \alpha_2 \in \mathbb{C}$ (see, e.g., [24, Proposition 4.10]). The multiple operator integral given by Definition 4.3.3 satisfies (4.3.4) for all bounded Borel functions $\varphi_1, \varphi_2 : \mathbb{R}^{n+1} \to \mathbb{C}$ for which it is defined, as it follows immediately from the definition.

The following result is established in [163, Lemma 3.2].

Proposition 4.3.7 *Let $n \in \mathbb{N}$ and $1 \leqslant p, p_j \leqslant \infty$, $j = 1,\ldots,n$, be such that $0 \leqslant \frac{1}{p} = \frac{1}{p_1} + \cdots + \frac{1}{p_n} \leqslant 1$. Let $X_j \in \mathcal{S}^{p_j}$. Let $A \in \mathcal{D}_{sa}$, let $\varphi : \mathbb{R}^{n+1} \to \mathbb{C}$ be a bounded Borel function, and let T_φ be the transformation associated with $A_0 = \ldots = A_n = A$ and φ according to Definition 4.3.3. The following assertions hold.*

(i) Let $T_\varphi : \mathcal{S}^{p_1} \times \cdots \times \mathcal{S}^{p_n} \to \mathcal{S}^p$ be bounded. If

$$\overline{\varphi}(\lambda_0, \lambda_1, \ldots, \lambda_n) := \overline{\varphi(\lambda_n, \lambda_{n-1}, \ldots, \lambda_0)},$$

then $T_{\overline{\varphi}} : \mathcal{S}^{p_n} \times \cdots \times \mathcal{S}^{p_1} \to \mathcal{S}^p$ is bounded and

$$\left\| T_\varphi \right\| = \left\| T_{\overline{\varphi}} \right\|.$$

(ii) Assume that $1 \leqslant p_0 \leqslant \infty$ and $\frac{1}{p_0} + \cdots + \frac{1}{p_n} = 1$. Assume also that $T_\varphi : \mathcal{S}^{p_1} \times \cdots \times \mathcal{S}^{p_n} \to \mathcal{S}^{\frac{p_0}{p_0-1}}$ exists and is bounded. Define

$$\varphi^*(\lambda_n, \lambda_0, \ldots, \lambda_{n-1}) := \varphi(\lambda_0, \ldots, \lambda_{n-1}, \lambda_n).$$

If $T_{\varphi^*} : \mathcal{S}^{p_0} \times \cdots \times \mathcal{S}^{p_{n-1}} \to \mathcal{S}^{\frac{p_n}{p_n-1}}$ exists and is bounded, then

$$\mathrm{Tr}\left(X_0\, T_\varphi(X_1,\ldots,X_n)\right) = \mathrm{Tr}\left(T_{\varphi^*}(X_0,\ldots,X_{n-1})\,X_n\right),$$

for $X_j \in \mathcal{S}^{p_j}$, $j = 0,\ldots,n$.

(iii) *Let $\varphi_1 : \mathbb{R}^{k+1} \to \mathbb{C}$ and $\varphi_2 : \mathbb{R}^{n-k+1} \to \mathbb{C}$ be bounded Borel functions such that the operators $T_{\varphi_1} : S^{p_1} \times \cdots \times S^{p_k} \to S^q$ and $T_{\varphi_2} : S^{p_{k+1}} \times \cdots \times S^{p_n} \to S^r$ exist and are bounded, where $\frac{1}{q} = \frac{1}{p_1} + \cdots + \frac{1}{p_k}$ and $\frac{1}{r} = \frac{1}{p_{k+1}} + \cdots + \frac{1}{p_n}$. If*

$$\psi(\lambda_0, \ldots, \lambda_n) := \varphi_1(\lambda_0, \ldots, \lambda_k) \cdot \varphi_2(\lambda_k, \ldots, \lambda_n),$$

then the operator T_ψ exists and is bounded on $S^{p_1} \times \cdots \times S^{p_n}$ and

$$T_\psi(X_1, \ldots, X_n) = T_{\varphi_1}(X_1, \ldots, X_k) \cdot T_{\varphi_2}(X_{k+1}, \ldots, X_n),$$

for $X_j \in S^{p_j}$, $j = 1, \ldots, n$.

(iv) *Let $\varphi_1 : \mathbb{R}^{k+1} \to \mathbb{C}$ and $\varphi_2 : \mathbb{R}^{n-k+2} \to \mathbb{C}$ be bounded Borel functions such that $T_{\varphi_1} : S^{p_1} \times \cdots \times S^{p_k} \to S^q$ and $T_{\varphi_2} : S^q \times S^{p_{k+1}} \times \cdots \times S^{p_n} \to S^r$ exist and are bounded, where $\frac{1}{q} = \frac{1}{p_1} + \cdots + \frac{1}{p_k}$ and $\frac{1}{r} = \frac{1}{q} + \frac{1}{p_{k+1}} + \cdots + \frac{1}{p_n}$. If*

$$\psi(\lambda_0, \ldots, \lambda_n) := \varphi_1(\lambda_0, \ldots, \lambda_k) \cdot \varphi_2(\lambda_0, \lambda_k, \ldots, \lambda_n),$$

then the operator T_ψ exists and is bounded on $S^{p_1} \times \cdots \times S^{p_n}$ and

$$T_\psi(X_1, \ldots, X_n) = T_{\varphi_2}\big(T_{\varphi_1}(X_1, \ldots, X_k), X_{k+1}, \ldots, X_n\big),$$

for $X_j \in S^{p_j}$, $j = 1, \ldots, n$.

Proof Given $m, N \in \mathbb{N}$ and a bounded Borel function $\varphi : \mathbb{R}^{n+1} \to \mathbb{C}$, we denote

$$T_\varphi^{m,N}(X_1, \ldots, X_n) \tag{4.3.5}$$

$$:= \sum_{\substack{|l_j| \leqslant N \\ 0 \leqslant j \leqslant n}} \varphi\left(\frac{l_0}{m}, \ldots, \frac{l_n}{m}\right) E\left(\left[\frac{l_0}{m}, \frac{l_0}{m} + 1\right)\right) X_1 \ldots X_n E\left(\left[\frac{l_n}{m}, \frac{l_n}{m} + 1\right)\right),$$

where $X_j \in S^{p_j}$, $j = 0, \ldots, n$. Note that by the definition of T_φ,

$$T_\varphi(X_1, \ldots, X_n) = \lim_{m \to \infty} \lim_{N \to \infty} T_\varphi^{m,N}(X_1, \ldots, X_n). \tag{4.3.6}$$

(i) By taking the adjoint in (4.3.5) we obtain

$$\left(T_\varphi^{m,N}(X_1, \ldots, X_n)\right)^* = T_{\overline{\varphi}}^{m,N}(X_n^*, \ldots, X_1^*). \tag{4.3.7}$$

By (4.3.6),

$$T_{\overline{\varphi}}(X_n^*, \ldots, X_1^*) = \lim_{m \to \infty} \lim_{N \to \infty} T_{\overline{\varphi}}^{m,N}(X_n^*, \ldots, X_1^*) \tag{4.3.8}$$

Combining (4.3.7)–(4.3.8) implies

$$\left(T_\varphi(X_1, \ldots, X_n)\right)^* = T_{\overline{\varphi}}(X_n^*, \ldots, X_1^*).$$

Hence, $\|T_\varphi : \mathcal{S}^{p_1} \times \cdots \times \mathcal{S}^{p_n} \to \mathcal{S}^p\| = \|T_{\overline{\varphi}} : \mathcal{S}^{p_n} \times \cdots \times \mathcal{S}^{p_1} \to \mathcal{S}^p\|$.

(ii) By cyclicity of the trace,

$$\mathrm{Tr}\left(\varphi\left(\frac{l_0}{m}, \ldots, \frac{l_n}{m}\right) X_0 E_{l_0,m} X_1 E_{l_1,m} X_2 \ldots E_{l_{n-1},m} X_n E_{l_n,m}\right)$$

$$= \mathrm{Tr}\left(\varphi\left(\frac{l_0}{m}, \ldots, \frac{l_n}{m}\right) E_{l_n,m} X_0 E_{l_0,m} X_1 E_{l_1,m} X_2 \ldots E_{l_{n-1},m} X_n\right).$$

Hence,

$$\mathrm{Tr}\left(X_0 T_\varphi^{m,N}(X_1, \ldots, X_n)\right) = \mathrm{Tr}\left(T_{\varphi^*}^{m,N}(X_0, \ldots, X_{n-1}) X_n\right). \qquad (4.3.9)$$

Combining (4.3.6) and (4.3.9) completes the proof of the claim.

(iii) Immediately from (4.3.5) we obtain

$$T_\psi^{m,N}(X_1, \ldots, X_n) = T_{\varphi_1}^{m,N}(X_1, \ldots, X_k) \cdot T_{\varphi_2}^{m,N}(X_{k+1}, \ldots, X_n).$$
$$(4.3.10)$$

Passing to the limit in (4.3.10) and applying (4.3.6) proves the claim.

(iv) Immediately from (4.3.5) we obtain

$$T_\psi^{m,N}(X_1, \ldots, X_n) = T_{\varphi_2}^{m,N}\left(T_{\varphi_1}^{m,N}(X_1, \ldots, X_k), X_{k+1}, \ldots, X_n\right).$$
$$(4.3.11)$$

Passing to the limit in (4.3.11) and applying (4.3.6) proves the claim. □

The proof of Proposition 4.3.7 is an example of a general approach to deriving results on multiple operator integrals given by Definition 4.3.3: prove results for an operator A whose spectrum is a finite subset of $\mathbb{Z}$ (by working with a finite sum $T_\varphi^{m,N}$ defined in (4.3.5)) and then transfer the results to an operator A with an arbitrary spectrum by approximations. It is also convenient to establish results for a multiple operator integral associated with $A_0 = \cdots = A_n$ and then transfer the results to a multiple operator integral associated with distinct $A_0, \ldots, A_n$ by adopting a trick with tensor products illustrated in Proposition 4.1.8. This approach is utilized in [163, 164, 191] and demonstrated in the proof of Theorem 4.3.10 below.

Norm Estimates

Immediately from Definition 4.3.1 and Hölder's inequality, we have the following estimate for a multiple operator integral.

Theorem 4.3.8 *Let* $n \in \mathbb{N}$, $A_1, \ldots, A_{n+1} \in \mathcal{D}_{sa}$, $\varphi \in \mathfrak{A}_n$, *and* $0 \leqslant \frac{1}{p} = \frac{1}{p_1} + \cdots + \frac{1}{p_n} \leqslant 1$. *Then,*

$$\left\| T_\varphi^{A_1,\ldots,A_{n+1}} (X_1, \ldots, X_n) \right\|_p \leqslant \|\varphi\|_{\mathfrak{A}_n} \|X_1\|_{p_1} \cdots \|X_n\|_{p_n} \qquad (4.3.12)$$

for all $X_j \in S^{p_j}$ *(or* $X_j \in \mathcal{B}(\mathcal{H})$ *if* $p_j = \infty$*),* $j = 1, \ldots, n$.

An estimate with a norm of the symbol φ smaller than $\|\varphi\|_{\mathfrak{A}_n}$ is obtained in [163] for the multiple operator integral given by (4.3.3) via a subtle analysis. We state the result and outline the scheme of its proof below.

The largest set of symbols for which the aforementioned result holds is described via polynomial integral momenta introduced in [163]. Let S_n be the simplex

$$S_n = \left\{ (s_0, \ldots, s_n) \in \mathbb{R}^{n+1} : \sum_{j=0}^n s_j = 1, \ s_j \geqslant 0, \ j = 0, \ldots, n \right\} \qquad (4.3.13)$$

equipped with the finite measure $d\sigma_n$ defined by

$$\int_{S_n} \phi(s_0, \ldots, s_n) \, d\sigma_n = \int_{R_n} \phi \left(s_0, \ldots, s_{n-1}, 1 - \sum_{j=0}^{n-1} s_j \right) dv_n.$$

for every continuous function $\phi : \mathbb{R}^{n+1} \to \mathbb{C}$, where

$$R_n = \left\{ (s_0, \ldots, s_{n-1}) \in \mathbb{R}^n : \sum_{j=0}^{n-1} s_j \leqslant 1, \ s_j \geqslant 0, \ j = 0, \ldots, n-1 \right\}$$

and dv_n is the Lebesgue measure on $\mathbb{R}^n$. Let $\mathcal{P}_n$ denote the set of polynomials with real coefficients of $n + 1$ variables. Given $h \in C_b(\mathbb{R})$ and $\wp \in \mathcal{P}_n$, we call the function $\phi_{n,h,\wp}$ defined by

$$\phi_{n,h,\wp}(\lambda_0, \ldots, \lambda_n) = \int_{S_n} \wp(s_0, \ldots, s_n) h\left(\sum_{j=0}^n s_j \lambda_j \right) d\sigma_n \qquad (4.3.14)$$

a *polynomial integral momentum.* The function ϕ is continuous because h is.

The divided difference is an example of a polynomial integral momentum, as it follows from standard properties of divided differences (see, e.g., [163, Lemma 5.1], where a detailed comparison of these definitions is presented).

Proposition 4.3.9 *For* $n \in \mathbb{N}$, $f \in C^n(\mathbb{R})$,

$$f^{[n]} = \phi_{n, f^{(n)}, 1},$$

where $\phi_{n, f^{(n)}, 1}$ *is given by* (4.3.14).

The following estimate is established in [163, Theorem 5.6].

Theorem 4.3.10 *Let $n \in \mathbb{N}$, $\wp \in \mathcal{P}_n$, $h \in C_b(\mathbb{R})$, and let $\phi_{n,h,\wp}$ be the polynomial integral momentum defined by (4.3.14). Let $1 < p, p_j < \infty$, $j = 1, \ldots, n$, be such that $0 < \frac{1}{p} = \frac{1}{p_1} + \cdots + \frac{1}{p_n} < 1$. Let $A_j \in \mathcal{D}_{sa}$, $j = 0, \ldots, n$ and let $T^{A_0,\ldots,A_n}_{\phi_{n,h,\wp}}$ be the transformation given by Definition 4.3.3. Then,*

$$T_{\phi_{n,h,\wp}} \in \mathcal{B}_n(S^{p_1} \times \cdots \times S^{p_n}, S^p)$$

and there exists $c_{\wp,p_1,\ldots,p_n} > 0$ such that

$$\|T^{A_0,\ldots,A_n}_{\phi_{n,h,\wp}}\| \leqslant c_{\wp,p_1,\ldots,p_n} \|h\|_\infty. \tag{4.3.15}$$

Proof of Theorem 4.3.10 The result is proved by induction on n.

Assume first that $A_0 = \cdots = A_n =: A$ and $\sigma(A)$ is a finite subset of $\mathbb{Z}$. In this case $T_{\phi_{n,h,\wp}} := T^{A,\ldots,A}_{\phi_{n,h,\wp}}$ is a finite sum.

(1) Base of induction: $n = 1$.

Let $2 < p_1 < \infty$ be sufficiently large. It is proved in [163, Theorem 5.6] (see also [163, Lemma 4.6]) that if $2 < q < p_1$ satisfies $\frac{1}{p_1} + \frac{1}{q} = \frac{1}{2}$, then there exists $\tilde{c}_{\wp,p_1} > 0$ such that

$$\|T_{\phi_{1,h,\wp}} : S^{p_1} \to S^{p_1}\| \leqslant \tilde{c}_{\wp,p_1}\left(1 + \|T_{\phi_{1,h,\wp}} : S^q \to S^q\|\right). \tag{4.3.16}$$

We present two key technical ideas of the proof, which are also used in the proof of the induction step below. They consist in decomposing the whole multiple operator integral into simpler components that can be handled by means of harmonic analysis.

Note that

$$T_\phi(X) = T_\phi(\mathcal{T}_u(X)) + T_\phi(X_d) + T_\phi(\mathcal{T}_l(X)) = \sum_{\lambda_0 < \lambda_1} \phi(\lambda_0, \lambda_1)E(\{\lambda_0\})XE(\{\lambda_1\})$$

$$+ \sum_\lambda \phi(\lambda, \lambda)E(\{\lambda\})XE(\{\lambda\})$$

$$+ \sum_{\lambda_0 > \lambda_1} \phi(\lambda_0, \lambda_1)E(\{\lambda_0\})XE(\{\lambda_1\}),$$

where $\mathcal{T}_u$ and $\mathcal{T}_l$ are respectively upper and lower triangular truncations, which are bounded operators on S^p, $1 < p < \infty$ [85, 91], and X_d is the part of X diagonal with respect to the spectral measure E. Thus, estimating the whole multiple operator integral can be reduced to estimating it on truncated perturbations.

It is proved in [159, Lemma 6] using Fourier inversion that there exists a function $g : \mathbb{R} \to \mathbb{C}$ satisfying

$$\int_{\mathbb{R}} |s|^n |g(s)| \, ds < \infty, \quad n \in \mathbb{N} \cup \{0\},$$

and

$$\frac{l_1 - l_2}{l_1 - l_0} = \frac{1}{\sqrt{2\pi}} \int_{\mathbb{R}} g(s)(l_1 - l_2)^{is}(l_1 - l_0)^{-is} \, ds \qquad (4.3.17)$$

for

$$\frac{|l_1 - l_2|}{|l_1 - l_0|} \leqslant 1.$$

Define

$$X_{s,1} = \sum_{l_0 < l_1} (l_1 - l_0)^{is} E_{l_0} X_1 E_{l_1} \quad \text{and} \quad X_{s,2} = \sum_{l_2 < l_1} (l_1 - l_2)^{is} E_{l_2} X_2 E_{l_1}.$$

It is proved in [159, Lemma 4.5] (see also [91, Proposition 5.4.8]) that there exist $\gamma_{p_i} > 0, i = 1, 2$, such that

$$\|X_{s,i}\| \leqslant \gamma_{p_i}(1 + |s|)\|X\|_{p_i}. \qquad (4.3.18)$$

The bound (4.3.18) is employed multiple times in the proof.

There exists $\theta \in (0, 1)$ such that $\frac{1}{q} = \frac{1-\theta}{2} + \frac{\theta}{p_1}$. By the complex interpolation method (see, e.g., [30]),

$$\left\|T_{\phi_{1,h,\wp}} : \mathcal{S}^q \to \mathcal{S}^q\right\| \leqslant \left\|T_{\phi_{1,h,\wp}} : \mathcal{S}^2 \to \mathcal{S}^2\right\|^{1-\theta} \left\|T_{\phi_{1,h,\wp}} : \mathcal{S}^{p_1} \to \mathcal{S}^{p_1}\right\|^{\theta}. \qquad (4.3.19)$$

By (3.2.2),

$$\left\|T_{\phi_{1,h,\wp}} : \mathcal{S}^2 \to \mathcal{S}^2\right\|^{1-\theta} \leqslant \|\phi_{1,h,\wp}\|_{\infty}^{1-\theta}. \qquad (4.3.20)$$

Combining (4.3.16)–(4.3.20) implies that there exists $c_{\wp,p_1} > 0$ such that

$$\left\|T_{\phi_{1,h,\wp}} : \mathcal{S}^{p_1} \to \mathcal{S}^{p_1}\right\| \leqslant c_{\wp,p_1}\|h\|_{\infty},$$

By duality and Proposition 4.3.7(i) and then by interpolation, the latter bound implies (4.3.15) with $n = 1$ for all values of $p_1 \in (1, \infty)$.

(2) Induction step: show that

$$\|T_{\phi_{n-1,h,\wp}}\| \leqslant c_{\wp,p_1,\dots,p_{n-1}} \|h\|_\infty$$

implies (4.3.15).

The proof involves reduction of order of the polynomial integral momentum, which exists only for a certain subset of the variables $(\lambda_0, \dots, \lambda_n)$ considered below. Let $\lambda_0 \leqslant \lambda_2 \leqslant \lambda_1$ with $\lambda_0 \neq \lambda_1$. One of the key technical steps in the proof is the decomposition

$$\int_0^\kappa t^m \, dt \int_0^t s^k \, h(\kappa\lambda_2 + (\lambda_0 - \lambda_2)t + (\lambda_1 - \lambda_0)s) \, ds$$

$$= \int_0^\kappa \tilde{\wp}_1 \left(\frac{\lambda_0 - \lambda_2}{\lambda_0 - \lambda_1}, \kappa, \theta \right) h(\kappa\lambda_2 + (\lambda_0 - \lambda_2)\theta) \, d\theta \qquad (4.3.21)$$

$$+ \int_0^\kappa \tilde{\wp}_2 \left(\frac{\lambda_0 - \lambda_2}{\lambda_0 - \lambda_1}, \kappa, \sigma \right) h(\kappa\lambda_2 + (\lambda_1 - \lambda_2)\sigma) \, d\sigma$$

derived in [163, Lemma 5.9] for $\kappa > 0$, where $\tilde{\wp}_1$ and $\tilde{\wp}_2$ are the polynomials depending on m and k but not on h and given by

$$\tilde{\wp}_1(\zeta, \kappa, \theta) = \zeta^{k+1} \int_\theta^\kappa (t - \theta)^k t^m \, dt,$$

$$\tilde{\wp}_2(\zeta, \kappa, \sigma) = (1 - \zeta) \int_\sigma^\kappa ((1 - \zeta)\sigma + \zeta t)^k t^m \, dt.$$

The decomposition (4.3.21) is used in [163, Lemma 5.8] to obtain the following order reduction of the polynomial integral momentum:

$$\phi_{n,h,\wp}(\lambda_0, \lambda_1, \lambda_2, \lambda_3, \dots, \lambda_n) = \psi_{n-1,h,\wp_1} \left(\frac{\lambda_0 - \lambda_2}{\lambda_0 - \lambda_1}, \lambda_0, \lambda_2, \lambda_3, \dots, \lambda_n \right)$$

$$+ \psi_{n-1,h,\wp_2} \left(\frac{\lambda_0 - \lambda_2}{\lambda_0 - \lambda_1}, \lambda_1, \lambda_2, \lambda_3, \dots, \lambda_n \right),$$

$$(4.3.22)$$

where $\wp_1, \wp_2 \in \mathcal{P}_n$ depend only on $\wp$ and

$$\psi_{n-1,h,\wp_i}(\lambda, \lambda_{i-1}, \lambda_2, \dots, \lambda_n)$$

$$= \int_{S_{n-1}} \wp_i(\lambda, s_0, \dots, s_{n-1}) h\left(\sum_{j=0}^{n-1} s_j \lambda_j \right) d\sigma_{n-1}, \quad i = 1, 2,$$

with S_n defined in (4.3.13). The polynomial $\wp_i(\zeta, s_0, \dots, s_{n-1})$, $i = 1, 2$, is a linear combination of the polynomials $\breve{\wp}(\zeta, s_0, \dots, s_{n-1}) = (1 - \zeta)^m \breve{\wp}_r(s_0, \dots, s_{n-1})$, where $\breve{\wp}_r \in \mathcal{P}_{n-1}$ and $m \in \mathbb{N} \cup \{0\}$.

Consider the transformations

$$T_{\phi_{n,h,\wp}}^{(0,2,1)}(X_1,\ldots,X_n) := \sum_{l_0 \leqslant l_2 < l_1} \phi_{n,h,\wp}(l_0,\ldots,l_n) E_{l_0} X_1 E_{l_1} X_2 \ldots X_n E_{l_n},$$

$$Q_{n-1,h,\wp_i}^{(0,2,1)}(X_1,\ldots,X_n) := \sum_{l_0 \leqslant l_2 < l_1} \psi_{n-1,h,\wp_i}(l_0,\ldots,l_n)$$

$$\times E_{l_0} X_1 E_{l_1} X_2 \ldots X_n E_{l_n} i = 1, 2.$$

where $E_{l_j} = E_A([l_j, l_j + 1))$, $l_j \in \mathbb{Z}$ (recall the assumption $\sigma(A) \subset \mathbb{Z}$), $j = 0, \ldots, n$, and $\wp_1, \wp_2$ satisfy (4.3.22). It follows from (4.3.4) and (4.3.22) that

$$T_{\phi_{n,h,\wp}}^{(0,2,1)} = Q_{n-1,h,\wp_1}^{(0,2,1)} + Q_{n-1,h,\wp_2}^{(0,2,1)}.$$

Thanks to the latter representation, property (4.3.4), and structure of the polynomials $\wp_i$, $i = 1, 2$, in order to prove

$$\|T_{\phi_{n,h,\wp}}^{(0,2,1)}\| \leqslant c_{\wp,p_1,\ldots,p_n} \|h\|_\infty \tag{4.3.23}$$

it suffices to prove

$$\|Q_{n-1,h,\breve{\wp}}^{(0,2,1)}\| \leqslant c_{\wp,p_1,\ldots,p_n} \|h\|_\infty,$$

where we use the same notation for various constants depending on the same parameters.

By Proposition 4.3.7(iii) and (iv),

$$Q_{n-1,h,\breve{\wp}}^{(0,2,1)}(X_1,\ldots,X_n) = \int_{\mathbb{R}} g(s) X_{-ms,1} T_{\phi_{n-1,h,\wp r}}(X_{ms,2}, X_3, \ldots, X_n)\, ds.$$

By the triangle and Hölder inequalities,

$$\|Q_{n-1,h,\breve{\wp}}^{(0,2,1)}(X_1,\ldots,X_n)\|_p$$

$$\leqslant \int_{\mathbb{R}} |g(s)| \, \|X_{-ms,1}\|_{p_1} \, \|T_{\phi_{n-1,h,\wp r}}(X_{ms,2}, X_3, \ldots, X_n)\|_{\frac{pp_1}{p_1-p}}\, ds. \tag{4.3.24}$$

By the induction hypothesis,

$$\|T_{\phi_{n-1,h,\wp r}}(X_{ms,2}, X_3, \ldots, X_n)\|_{\frac{pp_1}{p_1-p}} \leqslant \|X_{ms,2}\|_{p_2} \|X_3\|_{p_3} \ldots \|X_n\|_{p_n}. \tag{4.3.25}$$

Combining (4.3.24), (4.3.25), and (4.3.18) implies

$$\|Q_{n-1,h,\breve{\wp}}^{(0,2,1)}(X_1,\ldots,X_n)\|_p \leqslant \gamma_{p_1}\gamma_{p_2}\int_{\mathbb{R}}|g(s)|(1+|ms|)^2\,ds\,\|X_1\|_{p_1}\ldots\|X_n\|_{p_n}.$$

Hence, $Q_{n-1,h,\breve{\wp}}^{(0,2,1)} \in \mathcal{B}_n(S^{p_1}\times\cdots\times S^{p_n}, S^p)$ and, therefore, (4.3.23) holds.

Now we will demonstrate how the bound (4.3.15) can be derived from the bound (4.3.23). Let

$$D = \{(l_0,\ldots,l_n) \in \mathbb{Z}^{n+1} : l_0 = \cdots = l_n\},$$

$\epsilon \in \{-1,1\}^n$ and denote

$$K_\epsilon = \{(l_0,\ldots,l_n) \in \mathbb{Z}^{n+1} : l_{j-1} \leqslant l_j$$
$$\text{if } \epsilon_j = 1; \ l_{j-1} > l_j \text{ if } \epsilon_j = -1, j = 1,\ldots,n\}.$$

There is an index $0 \leqslant j_\epsilon \leqslant n$ such that

$$(l_0,\ldots,l_n) \in K_\epsilon \Rightarrow l_{j_\epsilon-1} \leqslant l_{j_\epsilon} \text{ and } l_{j_\epsilon} > l_{j_\epsilon+1}, \tag{4.3.26}$$

where the decrement and increment of the indices $j_\epsilon - 1$ and $j_\epsilon + 1$ are understood modulo n, that is, if $j = 0$, then $j - 1 = n$ and if $j = n$, then $j + 1 = 0$. By fixing j_ϵ we further split K_ϵ into subspaces $K_{\epsilon,i}$, $i = 0, 1$, where

$$K_{\epsilon,0} = \{(l_0,\ldots,l_n) \in K_\epsilon : l_{j_\epsilon-1} \leqslant l_{j_\epsilon+1}\} \text{ and}$$
$$K_{\epsilon,1} = \{(l_0,\ldots,l_n) \in K_\epsilon : l_{j_\epsilon-1} > l_{j_\epsilon+1}\}.$$

The space $\mathbb{Z}^{n+1} \setminus D$ splits into the disjoint union of 2^{n+1} sets $K_{\epsilon,i}$, $\epsilon \in \{-1,1\}^n$, $i = 0, 1$. Hence,

$$T_{\phi_{n,h,\wp}} = T_{\phi_{n,h,\wp}}^D + \sum_{\epsilon\in\{-1,1\}^n}\sum_{i=0,1} T_{\phi_{n,h,\wp}}^{\epsilon,i}, \tag{4.3.27}$$

where

$$T_{\phi_{n,h,\wp}}^D = \sum_{(l_0,\ldots,l_n)\in D} \phi_{n,h,\wp}(l_0,\ldots,l_n)E_{l_0}X_1\ldots X_n E_{l_n}$$

and

$$T_{\phi_{n,h,\wp}}^{\epsilon,i} = \sum_{(l_0,\ldots,l_n)\in K_{\epsilon,i}} \phi_{n,h,\wp}(l_0,\ldots,l_n)E_{l_0}X_1\ldots X_n E_{l_n}.$$

We fix $\epsilon \in \{-1, 1\}^n$ and $j_\epsilon \in \{0, 1, \ldots, n\}$ as in (4.3.26) and obtain

$$
(l_0, \ldots, l_n) \in K_{\epsilon, i} \Rightarrow
\begin{cases}
l_{j_\epsilon - 1} \leqslant l_{j_\epsilon + 1} < l_{j_\epsilon} & \text{if } i = 0 \\
l_{j_\epsilon + 1} < l_{j_\epsilon - 1} \leqslant l_{j_\epsilon} & \text{if } i = 1.
\end{cases}
$$

If $j_\epsilon = 1$ and $i = 0$, then $T_{\phi_n, h, \wp}^{\epsilon, i} = T_{\phi_n, h, \wp}^{(0,2,1)}$, which boundedness is proved in (4.3.23). By shifting and reversing the enumeration of variables as in Proposition 4.3.7(i) and (ii) we reduce estimating $T_{\phi_n, h, \wp}^{\epsilon, i}$ to estimating $T_{\phi_n, h, \wp}^{(0,2,1)}$. Therefore,

$$
\| T_{\phi_n, h, \wp}^{\epsilon, i} \| \leqslant c_{\wp, p_1, \ldots, p_n} \| h \|_\infty. \tag{4.3.28}
$$

We note that

$$
T_{\phi_n, h, \wp}^D (X_1, \ldots, X_n) = \int_{S_n} \wp(s_0, \ldots, s_n) \, d\sigma_n \sum_{l \in \sigma(A)} h(l) E_l \sum_{k \in \sigma(A)} E_k X_1 \ldots X_n E_k,
$$

so

$$
\| T_{\phi_n, h, \wp}^D (X_1, \ldots, X_n) \|_p \leqslant c_\wp \| h \|_\infty \left\| \sum_{k \in \sigma(A)} E_k X_1 \ldots X_n E_k \right\|_p. \tag{4.3.29}
$$

Consider the family of unitaries

$$
U_t = \sum_{l \in \sigma(A)} e^{2\pi i l t} E_l, \quad t \in [0, 1].
$$

Due to the orthogonality of the trigonometric functions,

$$
\sum_{k \in \sigma(A)} E_k X_1 \ldots X_n E_k = \int_0^1 \cdots \int_0^1 \prod_{j=1}^n U_{t_j}^* X_j U_{t_j} \, dt_1 \ldots dt_n. \tag{4.3.30}
$$

Applying the Hölder inequality in (4.3.30) and combining the outcome with (4.3.29) gives

$$
\| T_{\phi_n, h, \wp}^D \| \leqslant c_{\wp, p_1, \ldots, p_n} \| h \|_\infty. \tag{4.3.31}
$$

Combining (4.3.27), (4.3.28), and (4.3.31) proves (4.3.15) under the additional assumption $A_0 = \cdots = A_n = A$ and $\sigma(A)$ is a finite subset of $\mathbb{Z}$.

A careful investigation of the proofs of [163, Lemmas 3.3 and 5.5] shows that their results also hold for distinct $A_0, \ldots, A_n$. This immediately implies that to get (4.3.15) in the general case it suffices to prove (4.3.15) for $A_0, \ldots, A_n$ whose spectra are finite subsets of $\mathbb{Z}$. Below we reduce the case of distinct $A_0, \ldots, A_n$ to the case of identical $A_0 = \cdots = A_n$ that was handled above.

Assume that the spectra of $A_0, \ldots, A_n$ are finite subsets of $\mathbb{Z}$. Let $0 \leqslant k \leqslant n - 1$ and assume that the $(n-k)$ last self-adjoint operators $A_{k+1}, A_{k+2}, \ldots, A_n$ are equal. Let E_{ij} denote the elementary 2×2 matrix whose nonzero entry has indices (i, j). Let $\mathcal{X}_l = X_l \otimes E_{22}$ for $l = 1, \ldots, k - 1$, $\mathcal{X}_k = X_k \otimes E_{21}$, and $\mathcal{X}_l = X_l \otimes E_{11}$ for $l = k + 1, \ldots, n$. Then for every $l = 0, \ldots, k$, let $\mathcal{A}_l$ be the self-adjoint operator with the spectral measure $E_{A_{k+1}} \otimes E_{11} + E_{A_l} \otimes E_{22}$ and for every $l = k + 1, \ldots, n$, let $\mathcal{A}_l = \mathcal{A}_k$. Consider $X \in \mathcal{S}^{p'}$, with $\frac{1}{p} + \frac{1}{p'} = 1$, and let $\mathcal{X} = X \otimes E_{12}$. A straightforward calculation (see, e.g., the proof of [191, Theorem 3.3]) implies that

$$\mathrm{Tr}\left(T^{A_0, \ldots, A_n}_{\phi_{n,h,\wp}}(X_1, \ldots, X_n)X\right) = \mathrm{Tr}\left(T^{\mathcal{A}_0, \ldots, \mathcal{A}_n}_{\phi_{n,h,\wp}}(\mathcal{X}_1, \ldots, \mathcal{X}_n)\mathcal{X}\right).$$

Note that by construction, the $(n - k + 1)$ self-adjoint operators $\mathcal{A}_k, \mathcal{A}_{k+1}, \ldots, \mathcal{A}_n$ are equal. Further, $\|\mathcal{X}_l\|_p = \|X_l\|_p$ for every $l = 1, \ldots, n$ and $\|\mathcal{X}\|_{p'} = \|X\|_{p'}$. Using this process inductively for $k = n - 1, n - 2, \ldots, 0$, we obtain that to prove (4.3.15), it suffices to have it when the self-adjoint operators are all equal. This concludes the proof of the theorem. □

In view of its importance, we state a particular case of Theorem 4.3.10 for the polynomial integral momentum $\phi_{n,f^{(n)},1} = f^{[n]}$ (see Proposition 4.3.9).

Theorem 4.3.11 *Let $n \in \mathbb{N}$, $f \in C^n(\mathbb{R})$, $f^{(n)} \in C_b(\mathbb{R})$. Let $1 < p, p_j < \infty$, $j = 1, \ldots, n$, be such that $0 < \frac{1}{p} = \frac{1}{p_1} + \cdots + \frac{1}{p_n} < 1$. Let $A_j \in \mathcal{D}_{sa}$, $j = 0, \ldots, n$ and let $T^{A_0, \ldots, A_n}_{f^{[n]}}$ be the transformation given by Definition 4.3.3. Then,*

$$T^{A_0, \ldots, A_n}_{f^{[n]}} \in \mathcal{B}_n(\mathcal{S}^{p_1} \times \cdots \times \mathcal{S}^{p_n}, \mathcal{S}^p)$$

and there exists $c_{p_1, \ldots, p_n} > 0$ such that

$$\|T^{A_0, \ldots, A_n}_{f^{[n]}}\| \leqslant c_{p_1, \ldots, p_n} \|f^{(n)}\|_\infty.$$

Problem 4.3.12 Find an estimate for the constant $c_{\wp, p_1, \ldots, p_n}$ in (4.3.15).

Perturbation Formula

The following extension of (4.1.9) to the infinite dimensional case is obtained in [119, Lemma 3.10].

Theorem 4.3.13 *Let $1 < p < \infty$, $n \in \mathbb{N}$, $n \geqslant 2$, and $f \in C^n(\mathbb{R})$, $f^{(n-1)}, f^{(n)} \in C_b(\mathbb{R})$. Let $A_1, \ldots, A_{n-1}, A, B \in \mathcal{D}_{sa}$ with $B - A \in \mathcal{S}^p$, let $X_1, \ldots, X_{n-1} \in \mathcal{S}^p$. Then, for every $i = 1, \ldots, n$,*

$$T^{A_1, \ldots, A_{i-1}, A, A_i, \ldots, A_{n-1}}_{f^{[n-1]}}(X_1, \ldots, X_{n-1}) - T^{A_1, \ldots, A_{i-1}, B, A_i, \ldots, A_{n-1}}_{f^{[n-1]}}(X_1, \ldots, X_{n-1})$$

$$= T^{A_1, \ldots, A_{i-1}, A, B, A_i, \ldots, A_{n-1}}_{f^{[n]}}(X_1, \ldots, X_{i-1}, A - B, X_i, \ldots, X_{n-1}).$$

The assumption of $f^{(n)}$ be continuous in Theorem 4.3.13 is eliminated in [52] for the transformation Γ given by Definition 4.2.1.

Theorem 4.3.13 has an analog for perturbations in $\mathcal{S}^1$ (or $\mathcal{B}(\mathcal{H})$) and a smaller set of functions f. Below we state the result of [119, Lemma 4.2] extending [146, Lemma 5.4] from the case $n = 2$ to a general n.

Theorem 4.3.14 *Let* $n \in \mathbb{N}$, $n \geqslant 2$, *and* $f \in B_{\infty 1}^{n-1}(\mathbb{R}) \cap B_{\infty 1}^n(\mathbb{R})$. *Let* $A_1, \ldots, A_{n-1}, A, B \in \mathcal{B}_{sa}(\mathcal{H})$ *with* $B - A \in \mathcal{S}^1$ *(or* $\mathcal{B}(\mathcal{H})$*), let* $X_1, \ldots, X_{n-1} \in \mathcal{S}^1$ *(or* $\mathcal{B}(\mathcal{H})$*). Then, for every* $i = 1, \ldots, n$,

$$T_{f^{[n-1]}}^{A_1,\ldots,A_{i-1},A,A_i,\ldots,A_{n-1}}(X_1,\ldots,X_{n-1}) - T_{f^{[n-1]}}^{A_1,\ldots,A_{i-1},B,A_i,\ldots,A_{n-1}}(X_1,\ldots,X_{n-1})$$

$$= T_{f^{[n]}}^{A_1,\ldots,A_{i-1},A,B,A_i,\ldots,A_{n-1}}(X_1,\ldots,X_{i-1},A-B,X_i,\ldots,X_{n-1}).$$

Continuity

The following continuity properties of multiple operator integrals can be established completely analogously to the properties established in Proposition 3.3.9. Recall that the class $\mathfrak{C}_n$ is defined in Sect. 4.3.2.

Proposition 4.3.15 *Let* $1 \leqslant p, p_1, \ldots, p_n \leqslant \infty$ *satisfy* $\frac{1}{p} = \frac{1}{p_1} + \cdots + \frac{1}{p_n}$. *Let* $\{A_{i,m}\}_{m=1}^\infty \subset \mathcal{D}_{sa}$ *converge to* $A_i \in \mathcal{D}_{sa}$, $i = 1, \ldots, n+1$, *in the strong resolvent sense and let* $\{X_{i,m}\}_{m=1}^\infty \subset \mathcal{S}^{p_i}$ *(*$\mathcal{B}(\mathcal{H})$*, respectively) converge to* $X_i \in \mathcal{S}^{p_i}$ *(*$\mathcal{B}(\mathcal{H})$*, respectively),* $i = 1, \ldots, n$. *The following assertions hold for multiple operator integrals given by Definition 4.3.1.*

(i) *Let* $\varphi \in \mathfrak{C}_n$. *If* $X_i \in \mathcal{S}^{p_i}$, $1 \leqslant p < \infty$, *then*

$$\lim_{m \to \infty} \|T_\varphi^{A_{1,m},\ldots,A_{n+1,m}}(X_1,\ldots,X_n) - T_\varphi^{A_1,\ldots,A_{n+1}}(X_1,\ldots,X_n)\|_p = 0;$$

(4.3.32)

if $X_i \in \mathcal{B}(\mathcal{H})$, *then*

$$sot\text{-}\lim_{m \to \infty}(T_\varphi^{A_{1,m},\ldots,A_{n+1,m}}(X_1,\ldots,X_n) - T_\varphi^{A_1,\ldots,A_{n+1}}(X_1,\ldots,X_n)) = 0.$$

(4.3.33)

Moreover, if $\{A_{i,m}\}_{m=1}^\infty$ *converges to* A_i, $i = 1, \ldots, n+1$, *in the operator norm, then the strong operator topology convergence in (4.3.33) can be replaced with the operator norm convergence.*

(ii) *For every* $\varphi \in \mathfrak{A}_n$,

$$\lim_{m \to \infty} \|T_\varphi^{A_1,\ldots,A_{n+1}}(X_{1,m},\ldots,X_{n,m}) - T_\varphi^{A_1,\ldots,A_{n+1}}(X_1,\ldots,X_n)\|_p = 0.$$

(iii) Let $\{\varphi_k\}_{k=1}^{\infty} \subset \mathfrak{A}_n$ converge to φ in the norm $\|\cdot\|_{\mathfrak{A}_n}$. Then,

$$\lim_{k \to \infty} \|T_{\varphi_k}^{A_1,\ldots,A_{n+1}}(X_1,\ldots,X_n) - T_{\varphi}^{A_1,\ldots,A_{n+1}}(X_1,\ldots,X_n)\|_p = 0.$$

The result of Proposition 4.3.15(i) holds for a broader set of symbols φ when $p = 2$. The following fact is established in [55, Proposition 3.1].

Proposition 4.3.16 *Let $\{A_{i,m}\}_{m=1}^{\infty} \subset \mathcal{D}_{sa}$ converge to $A_i \in \mathcal{D}_{sa}$, $i = 1,\ldots,n+1$, in the strong resolvent sense and let $X_1,\ldots,X_n \in \mathcal{S}^2$. Then, (4.3.32) with $p = p_1 = \cdots = p_n = 2$ holds for every $\varphi \in C_b(\mathbb{R}^{n+1})$.*

With help of Proposition 4.3.15 one can transfer results from simpler $A_1,\ldots,A_{n+1}$ or $X_1,\ldots,X_n$ or else φ to a general case by approximations.

4.3.4 Nonself-adjoint Case

By analogy with multiple operator integrals for self-adjoint operators defined in Sects. 4.3.1 and 4.3.2, one can define multiple operator integrals for unitary operators. The following properties of the transformations discussed in Sect. 3.3.5 extend to the unitary case: the algebraic properties, estimate (4.3.12) (see (4.3.34) below), perturbation formulas (3.3.7) and (3.3.8) (see [168, Lemma 2.4(i)]), continuity. A unitary analog of the estimate obtained in Theorem 3.3.4 is stated in Theorem 4.3.19 below.

Immediately from the definition of a multiple operator integral for unitary operators and Hölder inequality, we have the following estimate.

Theorem 4.3.17 *Let $A_1,\ldots,A_{n+1}$ be unitary operators, $\varphi \in \mathfrak{A}_n$, and let $0 \leqslant \frac{1}{p} = \frac{1}{p_1} + \cdots + \frac{1}{p_n} \leqslant 1$. Then,*

$$\left\|T_{\varphi}^{A_1,\ldots,A_{n+1}}(X_1,\ldots,X_n)\right\|_p \leqslant \|\varphi\|_{\mathfrak{A}_n}\|X_1\|_{p_1}\ldots\|X_n\|_{p_n}. \qquad (4.3.34)$$

for all $X_j \in \mathcal{S}^{p_j}$ (or $X_j \in \mathcal{B}(\mathcal{H})$ if $p_j = \infty$), $j = 1,\ldots,n$.

It can be shown based on the results of [146] that

$$\left\|f^{[n]}\right\|_{\mathfrak{A}_n} \leqslant \mathrm{const}_n \|f\|_{B_{\infty 1}^n(\mathbb{T})} \qquad (4.3.35)$$

for $f \in B_{\infty 1}^n(\mathbb{T})$.

We start with an adjustment of Definition 4.3.3 to the case of unitary operators, which is due to [164, Definition 2.5].

Definition 4.3.18 Let $n, m \in \mathbb{N}$ and let $A_0,\ldots,A_n$ be unitary operators. Denote

$$z_{j,m} := e^{2\pi ij/m}, \qquad E_{j,m}^{(l)} := E_{A_l}([z_{j,m}, z_{j+1,m})),$$

for $j = 0, \ldots, m-1, l = 0, \ldots, n$. Let $1 \leqslant \alpha, \alpha_i \leqslant \infty$ for $i = 1, \ldots, n$ be such that $1/\alpha_1 + \cdots + 1/\alpha_n = 1/\alpha$. For ψ a bounded Borel function on $\mathbb{T}^{n+1}$, the mapping

$$T_\psi^{A_0, \ldots, A_n} : \mathcal{S}^{\alpha_1} \times \cdots \times \mathcal{S}^{\alpha_n} \to \mathcal{S}^{\alpha}$$

defined by

$$T_\psi^{A_0, \ldots, A_n}(X_1, \ldots, X_n)$$

$$:= \lim_{m \to \infty} \sum_{j_0, \ldots, j_n = 0}^{m-1} \psi(z_{j_0, m}, \ldots, z_{j_n, m}) E_{j_0, m}^{(0)} X_1 E_{j_1, m}^{(1)} X_2 \ldots X_n E_{j_n, m}^{(n)},$$

provided the limit exists for all $X_i \in \mathcal{S}^{\alpha_i}$, $i = 1, \ldots, n$, is called a multiple operator integral with symbol ψ.

Below we define algebraic analogs of polynomial integral momenta (4.3.14) considered in [191, Definition 2.3].

Given $n \in \mathbb{N}, l \in \{1, \ldots, n\}, d \in \{0, \ldots, n-l\}$, consider the polynomial

$$p_{l,d}(t_0, \ldots, t_{l-1}) := \sum_{\substack{d_0, \ldots, d_{l-1} \geqslant 0 \\ d_0 + \cdots + d_{l-1} = d}} a_{d, d_0, \ldots, d_{l-1}} t_0^{d_0} \cdots t_{l-1}^{d_{l-1}}.$$

and define the function

$$\varphi_{n-d, f^{(n)}, p_{l,d}}(\lambda_0, \ldots, \lambda_{l-1}, \lambda_{l+d}, \ldots, \lambda_n)$$

on $\mathbb{T}^{n-d+1}$ recursively as follows. For $d_0 \in \{0, \ldots, n-1\}$, set

$$\varphi_{n-d_0, f^{(n)}, t_0^{d_0}}(\lambda_0, \lambda_{d_0+1}, \ldots, \lambda_n) := d_0! \, f^{[n]}(\underbrace{\lambda_0, \ldots, \lambda_0}_{d_0+1}, \lambda_{d_0+1}, \ldots, \lambda_n);$$

for $d_0, \ldots, d_{l-1} \geqslant 0$ satisfying $d_0 + \cdots + d_{l-1} = d$, set

$$\varphi_{n-d, f^{(n)}, t_0^{d_0} \ldots t_{l-1}^{d_{l-1}}}(\lambda_0, \ldots, \lambda_{l-1}, \lambda_{l+d}, \ldots, \lambda_n)$$

$$:= d_0! \ldots d_{l-1}! \, f^{[n]}(\underbrace{\lambda_0, \ldots, \lambda_0}_{d_0+1}, \ldots, \underbrace{\lambda_{l-1}, \ldots, \lambda_{l-1}}_{d_{l-1}+1}, \lambda_{l+d}, \ldots, \lambda_n)$$

$$+ \sum_{(i_1, \ldots, i_{l-1}) \in I_l} (-1)^{d-d_0-i_1-\cdots-i_{l-1}+1} \binom{d_1}{i_1} \cdots \binom{d_{l-1}}{i_{l-1}}$$

$$\times \varphi_{n-d, f^{(n)}, t_0^{d-i_1-\cdots-i_{l-1}} t_1^{i_1} \ldots t_{l-1}^{i_{l-1}}}(\lambda_0, \ldots, \lambda_{l-1}, \lambda_{l+d}, \ldots, \lambda_n),$$

where $I_l = \left(\{0, \ldots, d_1\} \times \cdots \times \{0, \ldots, d_{l-1}\} \right) \setminus \{(d_1, \ldots, d_{l-1})\}$, and set

$$\varphi_{n-d, f^{(n)}, p_{l,d}}(\lambda_0, \ldots, \lambda_{l-1}, \lambda_{l+d}, \ldots, \lambda_n)$$

$$:= \sum_{\substack{d_0, \ldots, d_{l-1} \geqslant 0 \\ d_0 + \cdots + d_{l-1} = d}} a_{d, d_0, \ldots, d_{l-1}} \, \varphi_{n-d, f^{(n)}, t_0^{d_0} \ldots t_{l-1}^{d_{l-1}}}(\lambda_0, \ldots, \lambda_{l-1}, \lambda_{l+d}, \ldots, \lambda_n).$$

The result of Theorem 4.3.19 below is obtained in [164, Theorems 2.8 and 2.17] for a polynomial f and extended to functions with derivatives representable by absolutely convergent Fourier series in [191, Theorem 3.3].

Theorem 4.3.19 *Let* $n \in \mathbb{N}$, *let* $1 < p, p_j < \infty$, $j = 1, \ldots, n$, *satisfy* $0 < \frac{1}{p} = \frac{1}{p_1} + \cdots + \frac{1}{p_n} < 1$. *Let* $A_0, \ldots, A_n$ *be unitary operators on* $\mathcal{H}$. *Then, for every* $f \in C^n(\mathbb{T})$ *satisfying* $\sum_{j=-\infty}^{\infty} |j|^n |\mathcal{F}f(j)| < \infty$, *we have*

$$T_{f^{[n]}}^{A_0, \ldots, A_n} \in \mathcal{B}_n(S^{p_1} \times \cdots \times S^{p_n}, S^p)$$

and there exists $c_{p,n} > 0$ *such that*

$$\left\| T_{f^{[n]}}^{A_0, \ldots, A_n} \right\| \leqslant c_{p,n} \| f^{(n)} \|_\infty.$$

Remark 4.3.20 The bound of Theorem 4.3.19 extends to the bound

$$\left\| \Gamma^{A_0, \ldots, A_n}(f^{[n]}) : S^{p_1} \times \cdots \times S^{p_n} \to S^p \right\| \leqslant c_{p,n} \| f^{(n)} \|_\infty$$

for all functions $f \in C^n(\mathbb{T})$, where $\Gamma^{A_0, \ldots, A_n}$ is the extension of the transformation given by Definition 4.2.1 from the set $(S^2 \cap S^{p_1}) \times \cdots \times (S^2 \cap S^{p_n})$ to $S^{p_1} \times \cdots \times S^{p_n}$, by adjusting the methods of [52].

Proof of Theorem 4.3.19 The strategy of the proof is the same as for Theorem 4.3.10, but the technical realization is more subtle than in the self-adjoint case. We briefly outline the technical distinctions below.

Similarly to the self-adjoint case it suffices to prove the result for $A_0 = \cdots = A_n = A$ with spectrum contained in $\{e^{2\pi i j/m} : j = 0, \ldots, m-1\}$ with fixed $m \in \mathbb{N}$. Denote

$$T_{f^{[n]}} := T_{f^{[n]}}^{A, \ldots, A}, \quad E_j := E_A([z_{j,m}, z_{j+1,m})),$$

where $z_{j,m} = e^{2\pi i j/m}$. Given a Borel subset $B \subset \mathbb{T}^{n+1}$ and a Borel function ϕ on $\mathbb{T}^{n+1}$, denote

$$T_\phi^B(X_1, \ldots, X_n) := \sum_{(z_{j_0}, \ldots, z_{j_n}) \in B} \phi(z_{j_0}, \ldots, z_{j_n}) E_{j_0} X_1 E_{j_1} X_2 \ldots X_n E_{j_n},$$

for every $(X_1, \ldots, X_n) \in S^{p_1} \times \cdots \times S^{p_n}$.

By a modification of [159, Lemma 6], given $\delta > 0$, there exists a function $g_\delta : \mathbb{R} \to \mathbb{C}$ such that

$$\int_{\mathbb{R}} |s|^n |g_\delta(s)| \, ds < \infty, \quad n \in \mathbb{N} \cup \{0\},$$

and

$$\frac{\lambda_2 - \lambda_1}{\lambda_1 - \lambda_0} = \frac{1}{\sqrt{2\pi}} \left(\frac{|\lambda_1 - \lambda_0|}{\lambda_1 - \lambda_0} \right) \left(\frac{\lambda_2 - \lambda_1}{|\lambda_2 - \lambda_1|} \right) \int_{\mathbb{R}} g_\delta(s) \left(\frac{|\lambda_2 - \lambda_1|^{is}}{|\lambda_1 - \lambda_0|^{is}} \right) ds,$$

whenever

$$\frac{|\lambda_2 - \lambda_1|}{|\lambda_1 - \lambda_0|} \leqslant \delta,$$

which replaces (4.3.17) in the unitary case. The following analogs of (4.3.18) are established in [110, Theorem 3.4]. Let B, C be subsets of $\left\{ e^{2\pi i j/m} : j = 0, \ldots, m-1 \right\}$, let $r \in \mathbb{N}$, and denote

$$\Upsilon_r(X) := \sum_{z \in B, w \in C} \left(\frac{z - w}{|z - w|} \right)^r E(\{z\}) \, X \, E(\{w\})$$

$$\Upsilon_{-r}(X) := \sum_{z \in B, w \in C} \left(\frac{|z - w|}{z - w} \right)^r E(\{z\}) \, X \, E(\{w\})$$

$$\Gamma_s(X) := \sum_{z \in B, w \in C} |z - w|^{is} E(\{z\}) \, X \, E(\{w\}).$$

Then, there are constants $c_\alpha, c_{\alpha,r} > 0$ such that

$$\|\Upsilon_r(X)\|_\alpha \leqslant c_{\alpha,r} \|X\|_\alpha,$$

$$\|\Upsilon_{-r}(X)\|_\alpha \leqslant c_{\alpha,r} \|X\|_\alpha,$$

$$\|\Gamma_s(X)\|_\alpha \leqslant c_\alpha (1 + |s| + |s|^2) \|X\|_\alpha.$$

If f is analytic on the closed unit desk, then the decomposition (4.3.21) extends from the portion of the simplex on the real line determined by $0 \leqslant \lambda \leqslant \xi \leqslant \mu \leqslant 1$ to all $\lambda, \xi, \mu \in \mathbb{T}$ with $\lambda \neq \mu$ by uniqueness of meromorphic functions. Subsequently, we obtain (4.3.22) for $\lambda, \xi, \mu \in \mathbb{T}$ with $\lambda \neq \mu$. For a general function $f \in C^n(\mathbb{T})$ satisfying $\sum_{j=-\infty}^{\infty} |j|^n |\mathcal{F}f(j)| < \infty$, the representation (4.3.22) is replaced with algebraic counterparts given below. The algebraic approach is taken to avoid integration in (4.3.14) over a region containing the point 0, for which a trigonometric polynomial with negative powers of the variable is undefined.

Let $n \in \mathbb{N}$, $l \in \{1, \ldots, n\}$, $d \in \{0, \ldots, n-l\}$. Let $p_{l,d}$ be a polynomial of l variables

$$p_{l,d}(t_0, \ldots, t_{l-1}) := \sum_{\substack{d_0, \ldots, d_{l-1} \geqslant 0 \\ d_0 + \cdots + d_{l-1} = d}} a_{d, d_0, \ldots, d_{l-1}} \, t_0^{d_0} \ldots t_{l-1}^{d_{l-1}}.$$

If $n = 1$, then

$$\varphi_{1, f^{(n)}, t_0^d}(\lambda, \mu) = \left(\frac{\xi - \mu}{\lambda - \mu}\right)^n \varphi_{1, f^{(n)}, t_0^d}(\xi, \mu)$$

$$+ \sum_{j=0}^{n-1} \binom{n-1}{j} \left(\frac{\lambda - \xi}{\lambda - \mu}\right)^{j+1} \left(\frac{\xi - \mu}{\lambda - \mu}\right)^{n-1-j} \varphi_{1, f^{(n)}, t_0^d}(\lambda, \xi)$$

thanks to [191, Lemma 2.10]. If $n \geqslant 2$, then it is established in [191, Lemma 2.7] that there exist polynomials

$$q_{l,d,t_0,\ldots,t_{l-1}}(t_0, \ldots, t_{l-1}) = \sum_{\substack{k_1, \ldots, k_{l-1} \geqslant 0 \\ k_1 + \cdots + k_{l-1} = d+1}} b_{p_{l,d}, k_1, \ldots, k_{l-1}} \tilde{q}_{p_{l,d}, k_1, \ldots, k_{l-1}}(t_0) \, t_1^{k_1} \ldots t_{l-1}^{k_{l-1}}$$

and

$$r_{l,d,t_0,\ldots,t_{l-1}}(t_0, \ldots, t_{l-1}) = \sum_{\substack{k_1, \ldots, k_{l-1} \geqslant 0 \\ k_1 + \cdots + k_{l-1} = d+1}} c_{p_{l,d}, k_1, \ldots, k_{l-1}} \tilde{r}_{p_{l,d}, k_1, \ldots, k_{l-1}}(t_0) \, t_1^{k_1} \ldots t_{l-1}^{k_{l-1}},$$

where $\tilde{q}_{pl,d,k_1,\ldots,k_{l-1}}$ and $\tilde{r}_{pl,d,k_1,\ldots,k_{l-1}}$ are polynomials of one variable, such that for all $\lambda_0,\ldots,\lambda_{l-1},\lambda_{l+d},\ldots,\lambda_n \in \mathbb{T}$ with $\lambda_0 \neq \lambda_1$,

$$\varphi_{n-d,f^{(n)},pl,d}(\lambda_0,\ldots,\lambda_{l-1},\lambda_{l+d},\ldots,\lambda_n) \tag{4.3.36}$$

$$= \sum_{\substack{k_1,\ldots,k_{l-1}\geqslant 0 \\ k_1+\cdots+k_{l-1}=d+1}} b_{pl,d,k_1,\ldots,k_{l-1}}\,\tilde{q}_{pl,d,k_1,\ldots,k_{l-1}}\left(\frac{\lambda_0-\lambda_2}{\lambda_0-\lambda_1}\right)$$

$$\times \varphi_{n-d-d_1-1,f^{(n)},t_1^{k_1}\ldots t_{l-1}^{k_{l-1}}}(\lambda_0,\lambda_2,\ldots,\lambda_{l-1},\lambda_{l+d},\ldots,\lambda_n)$$

$$+ \sum_{\substack{k_1,\ldots,k_{l-1}\geqslant 0 \\ k_1+\cdots+k_{l-1}=d+1}} c_{pl,d,k_1,\ldots,k_{l-1}}\,\tilde{r}_{pl,d,k_1,\ldots,k_{l-1}}\left(\frac{\lambda_0-\lambda_2}{\lambda_0-\lambda_1}\right)$$

$$\times \varphi_{n-d-d_0-1,f^{(n)},t_1^{k_1}\ldots t_{l-1}^{k_{l-1}}}(\lambda_1,\lambda_2,\ldots,\lambda_{l-1},\lambda_{l+d},\ldots,\lambda_n).$$

The analog of (4.3.27) requires more care in the unitary case. Denote

$$Q_k^{(n)} := \left\{z \in \mathbb{T} : \arg(z) \in \left[\frac{2\pi k}{n+2}, \frac{2\pi(k+1)}{n+2}\right)\right\}, \quad k=0,\ldots,n+1,$$

where $\arg(z)$ denotes the principal value of the argument of the complex number z. By additivity of the multiple operator integral,

$$T_\varphi = \sum_{k_0,k_1,\ldots,k_n \in \{0,1,\ldots,n+1\}} T_\varphi^{Q_{k_0}^{(n)} \times Q_{k_1}^{(n)} \times \cdots \times Q_{k_n}^{(n)}}.$$

There are two principal types of the set $Q_{k_0}^{(n)} \times Q_{k_1}^{(n)} \times \cdots \times Q_{k_n}^{(n)}$. One is when there exists an index $i \in \{0,1,\ldots,n\}$ such that[1] $|k_{i+1}-k_i| \geqslant 2$ and, hence, $|z-w| \geqslant c_n > 0$ for $z \in Q_{k_{i+1}}^{(n)}$, $w \in Q_{k_i}^{(n)}$, and the other is when $|k_{i+1}-k_i| \leqslant 1$ for all i. In the latter case,[2] there is $a \in (0,\pi]$ such that $\arg(z) \subseteq [a, a+\pi]$ whenever $z \in Q_{k_i}^{(n)}$, for each i. Thus, in this case we have the inequality $|z_{j_1}-z_{j_0}| > |z_{j_2}-z_{j_1}|$ whenever $z_{j_0}, z_{j_1}, z_{j_2} \in Q_{k_0}^{(n)} \cup Q_{k_1}^{(n)} \cup \cdots \cup Q_{k_n}^{(n)}$ and $j_0 \leqslant j_2 < j_1$. The subcase $k_0 = k_1 = \cdots = k_n$ is treated by decomposing the off-diagonal part of $Q_0^{(n)} \times \cdots \times Q_0^{(n)}$ into the disjoint union of 2^{n+1} sets $K_{\epsilon,i}$, $\epsilon \in \{-1,1\}^n$, $i = 0,1$, analogously to how it

---

[1] Here the increment and decrement of the index i is understood modulo n, that is, if $i = n$, then $i+1 = 0$ and if $i = 0$, then $i-1 = n$.

[2] If n is even, then a typical example is $\begin{cases} k_i = i, & i \leqslant \frac{n}{2} \\ k_i = n-i+1, & i > \frac{n}{2}. \end{cases}$

was done in the proof of Theorem 4.3.15. The other subcases are treated by a similar analysis. □

We note that multiple operator integrals analogous to those in Definition 4.3.1 were introduced for contractions in [147] and maximal dissipative operators in [6].

4.3.5 Change of Variables

In this section we give an example of the change of operator variables in the multiple operator integral. The respective transition from unbounded self-adjoint operators with non-Schatten-class difference to unitary operators with Schatten-class difference is applied in perturbation theory, for instance, in the proof of Theorem 5.5.5.

The following result is established in [166, Theorem 3.2] and [191, Theorem 5.2].

Theorem 4.3.21 *Let* $n \geq 2$. *Let* $H \in \mathcal{D}_{sa}$ *and* $V \in \mathcal{B}_{sa}(\mathcal{H})$. *Assume the condition*

$$(H + V - iI)^{-1} - (H - iI)^{-1} \in \mathcal{S}^n.$$

Define the unitary operators

$$U_0 := (H + iI)(II - iI)^{-1} \quad \text{and} \quad U_1 := (H + V + iI)(H + V - iI)^{-1}.$$

If

$$\varphi \in \left\{ f \in C^n(\mathbb{T}) : \sum_{j=-\infty}^{\infty} |\mathcal{F}f(j)| |j|^n < \infty \right\}$$

and

$$\psi(\lambda) = \varphi\left(\frac{\lambda + i}{\lambda - i}\right), \quad \lambda \in \mathbb{R},$$

then

$$T_{\varphi^{[n]}}^{U_0,\ldots,U_0}(U_1 - U_0, \ldots, U_1 - U_0) = \sum_{k=1}^{n} \sum_{0=i_0<\cdots<i_k=n} T_{\psi^{[k]}}^{H,\ldots,H}(V_{i_1-i_0}, \ldots, V_{i_k-i_{k-1}}),$$

where

$$V_p = \left((I - V(H + V - iI)^{-1})V(H - iI)^{-1}\right)^{p-1}(I - V(H + V - iI)^{-1})V,$$

for $p \in \mathbb{N}$.

4.4 Multiple Operator Integrals on Noncommutative and Weak L^p-Spaces

Let M be a semifinite von Neumann algebra equipped with a faithful normal semifinite trace τ. The theory of multiple operator integrals has been extended from Schatten classes and $\mathcal{B}(\mathcal{H})$ to the setting of noncommutative L^p-spaces and weak L^p-spaces. The approach based on factorization of the symbol is due to [24] and the approach not involving separation of variables is due to [163, 170].

4.4.1 Approach via Separation of Variables

The multiple operator integral given by Definition 4.3.1 was extended in [24] to perturbations in M or in a symmetrically normed ideal $\mathcal{I}$ of M with property (F).

It is proved in [24, Lemma 4.6] that if $X_1, \ldots, X_n \in M$ and at least one of X_j is an element of $\mathcal{I}$, then the transformation given by Definition 4.3.1 maps $(X_1, \ldots, X_n)$ to $\mathcal{I}$.

4.4.2 Approach Without Separation of Variables

The following definition extends Definition 4.3.3 to the noncommutative L^p-spaces, as it was done in [163].

Definition 4.4.1 Let $k \in \mathbb{N}$ and let $1 \leqslant p, p_j \leqslant \infty$, $j = 1, \ldots, k$, be such that $0 \leqslant \frac{1}{p} = \frac{1}{p_1} + \cdots + \frac{1}{p_k} \leqslant 1$. Let $\varphi : \mathbb{R}^{k+1} \to \mathbb{C}$ be a bounded Borel function and $A_0, \ldots, A_k$ be self-adjoint operators affiliated with M. Suppose that for all $X_j \in L^{p_j}(M, \tau)$, $j = 1, \ldots, k$, for every $m \in \mathbb{N}$ the series

$$S_{\varphi,m}(X_1, \ldots, X_k) := \sum_{l_0, \ldots, l_k \in \mathbb{Z}} \varphi\Big(\frac{l_0}{m}, \ldots, \frac{l_k}{m}\Big) E_{A_0}\Big(\Big[\frac{l_0}{m}, \frac{l_0+1}{m}\Big)\Big)$$

$$\times \prod_{j=1}^{k} \Big[X_j E_{A_j}\Big(\Big[\frac{l_j}{m}, \frac{l_j+1}{m}\Big)\Big)\Big]$$

converges in the norm of $L^p(M, \tau)$ and

$$(X_1, \ldots, X_k) \mapsto S_{\varphi,m}(X_1, \ldots, X_k), \quad m \in \mathbb{N},$$

is a sequence in $\mathcal{B}_k(L^{p_1}(M, \tau) \times \cdots \times L^{p_k}(M, \tau), L^p(M, \tau))$. If the sequence $\{S_{\varphi,m}\}_{m=1}^{\infty}$ converges pointwise to some multilinear operator $T_\varphi^{A_0, \ldots, A_k}$, then, according to the Banach-Steinhaus theorem (see, e.g., [31]), $\{S_{\varphi,m}\}_{m=1}^{\infty}$ is uniformly bounded and $T_\varphi^{A_0, \ldots, A_k} \in \mathcal{B}_k(L^{p_1}(M, \tau) \times \cdots \times L^{p_k}(M, \tau), L^p(M, \tau))$. In this case, the operator $T_\varphi^{A_0, \ldots, A_k}$ is called a *multiple operator integral*.

Remark 4.4.2 It is proved in [163, Lemma 3.5] that the multiple operator integral T_φ discussed in Sect. 4.4.1 and the multiple operator integral given by Definition 4.4.1 coincide for $\varphi \in \mathfrak{C}_n$ on $\mathcal{M} \times \cdots \times \mathcal{M}$.

The following definition is an extension of Definition 4.4.1 to a Cartesian product of sums of L^p-spaces introduced in [170].

Definition 4.4.3 Let $k \in \mathbb{N}$ and let $1 \leqslant q, q_j, p, p_j \leqslant \infty$, $j = 1, \ldots, k$, be such that $0 \leqslant \frac{1}{q} = \frac{1}{q_1} + \cdots + \frac{1}{q_k} \leqslant 1$ and $0 \leqslant \frac{1}{p} = \frac{1}{p_1} + \cdots + \frac{1}{p_k} \leqslant 1$. Let $\varphi : \mathbb{R}^{k+1} \to \mathbb{C}$ be a bounded Borel function and $A_0, \ldots, A_k$ self-adjoint operators affiliated with $\mathcal{M}$. Suppose that for all $X_j \in (L^{q_j} + L^{p_j})(\mathcal{M}, \tau)$, $j = 1, \ldots, k$, for every $m \in \mathbb{N}$ the series

$$S_{\varphi,m}(X_1, \ldots, X_k) := \sum_{l_0, \ldots, l_k \in \mathbb{Z}} \varphi\left(\frac{l_0}{m}, \ldots, \frac{l_k}{m}\right) E_{A_0}\left(\left[\frac{l_0}{m}, \frac{l_0 + 1}{m}\right)\right)$$

$$\times \prod_{j=1}^{k} \left[X_j E_{A_j}\left(\left[\frac{l_j}{m}, \frac{l_j + 1}{m}\right)\right)\right]$$

converges in the norm of $(L^q + L^p)(\mathcal{M}, \tau)$ and

$$(X_1, \ldots, X_k) \mapsto S_{\varphi,m}(X_1, \ldots, X_k), \quad m \in \mathbb{N},$$

is a sequence in $\mathcal{B}_k((L^{q_1} + L^{p_1})(\mathcal{M}, \tau) \times \cdots \times (L^{q_k} + L^{p_k})(\mathcal{M}, \tau), (L^q + L^p)(\mathcal{M}, \tau))$. If the sequence $\{S_{\varphi,m}\}_{m=1}^{\infty}$ converges pointwise to some multilinear operator $T_\varphi^{A_0, \ldots, A_k}$, then, according to the Banach-Steinhaus theorem, $\{S_{\varphi,m}\}_{m=1}^{\infty}$ is uniformly bounded and $T_\varphi^{A_0, \ldots, A_k} \in \mathcal{B}_k((L^{q_1} + L^{p_1})(\mathcal{M}, \tau) \times \cdots \times (L^{q_k} + L^{p_k})(\mathcal{M}, \tau), (L^q + L^p)(\mathcal{M}, \tau))$. In this case, the operator $T_\varphi^{A_0, \ldots, A_k}$ is also called a *multiple operator integral*.

4.4.3 Properties of Multiple Operator Integrals on $L^{p,\infty}(\mathcal{M}, \tau)$

Algebraic Properties

The multiple operator integral discussed in Sect. 4.4.1 satisfies

$$T_{\alpha_1\varphi_1 + \alpha_2\varphi_2} = \alpha_1 T_{\varphi_1} + \alpha_2 T_{\varphi_2}, \tag{4.4.1}$$

for $\varphi_1, \varphi_2 \in \mathfrak{A}_n, \alpha_1, \alpha_2 \in \mathbb{C}$ (see, e.g., [24, Proposition 4.10]). The multiple operator integrals given by Definitions 4.4.1 and 4.4.3 satisfy (4.4.1) for all bounded Borel functions $\varphi_1, \varphi_2 : \mathbb{R}^{n+1} \to \mathbb{C}$ for which it is defined, as it follows immediately from the definition.

In the next two lemmas we collect simple properties of multiple operator integral, which proofs follow immediately from Definition 4.4.3.

Lemma 4.4.4 *Let* $k \in \mathbb{N}$ *and let* $A_0, \ldots, A_k$ *be self-adjoint operators affiliated with* $\mathcal{M}$. *Let* $k < r < q < \infty$ *and let* $\varphi : \mathbb{R}^{k+1} \to \mathbb{C}$ *be such that* $T_\varphi^{A_0,\ldots,A_k} \in \mathcal{B}_k((L^q + L^r)(\mathcal{M}, \tau)^{\times k}, (L^{\frac{q}{k}} + L^{\frac{r}{k}})(\mathcal{M}, \tau))$. *If* $A_0, A_1 \in \mathcal{M}$, *then for all* $X_1, \ldots, X_k \in (L^q + L^r)(\mathcal{M}, \tau)$ *the following assertions hold.*

(i) *If*

$$\psi_0(x_0, \ldots, x_k) := x_0 \varphi(x_0, \ldots, x_k),$$

then

$$T_{\psi_0}^{A_0, A_1, \ldots, A_k} \in \mathcal{B}_k((L^q + L^r)(\mathcal{M}, \tau)^{\times k}, (L^{\frac{q}{k}} + L^{\frac{r}{k}})(\mathcal{M}, \tau))$$

and

$$T_\varphi^{A_0, A_1, \ldots, A_k}(A_0 X_1, X_2, \ldots, X_k) = T_{\psi_0}^{A_0, A_1, \ldots, A_k}(X_1, X_2, \ldots, X_k).$$

(ii) *If*

$$\psi_1(x_0, \ldots, x_k) := x_1 \varphi(x_0, \ldots, x_k),$$

then

$$T_{\psi_1}^{A_0, A_1, \ldots, A_k} \in \mathcal{B}_k((L^q + L^r)(\mathcal{M}, \tau)^{\times k}, (L^{\frac{q}{k}} + L^{\frac{r}{k}})(\mathcal{M}, \tau))$$

and

$$T_\varphi^{A_0, A_1, \ldots, A_k}(X_1 A_1, X_2, \ldots, X_k) = T_{\psi_1}^{A_0, A_1, \ldots, A_k}(X_1, X_2, \ldots, X_k).$$

Proof (i) For $m \in \mathbb{N}$, directly from Definition 4.4.3, we obtain

$$S_{\varphi,m}(A_0 X_1, X_2, \ldots, X_k) = S_{\psi_0,m}(X_1, X_2, \ldots, X_k) + S_{\varphi,m}(A_{0,m} X_1, X_2, \ldots, X_k),$$

$$(4.4.2)$$

where

$$A_{0,m} := \sum_{l \in \mathbb{Z}} \left(A_0 - \frac{l}{m}\right) E_{A_0}\left(\left[\frac{l}{m}, \frac{l+1}{m}\right)\right).$$

Since $\{S_{\varphi,m}\}_{m \in \mathbb{N}}$ is a uniformly bounded sequence of operators and since

$$\|A_{0,m} X_1\|_{L^q + L^r} \to 0 \text{ as } m \to \infty,$$

it follows that

$$\|S_{\varphi,m}(A_{0,m}X_1, X_2, \ldots, X_k)\|_{L^{\frac{q}{k}}+L^{\frac{r}{k}}}$$

$$\leq \|S_{\varphi,m}\|_{(L^q+L^r)^{\times k}\to L^{\frac{q}{k}}+L^{\frac{r}{k}}} \|A_{0,m}X_1\|_{L^q+L^r} \prod_{j=2}^{k} \|X_j\|_{L^q+L^r} \to 0,$$

as $m \to \infty$. Since A_0 is bounded, it follows that $A_0X_1 \in (L^q + L^r)(\mathcal{M}, \tau)$. By Definition 4.4.3,

$$S_{\varphi,m}(A_0X_1, X_2, \ldots, X_k) \to T_{\varphi}^{A_0,A_1,\ldots,A_k}(A_0X_1, X_2, \ldots, X_k)$$

in $(L^{\frac{q}{k}} + L^{\frac{r}{k}})(\mathcal{M}, \tau)$ as $m \to \infty$. Taking the $(L^{\frac{q}{k}} + L^{\frac{r}{k}})(\mathcal{M}, \tau)$-limit of (4.4.2) completes the proof of (i).

The proof of (ii) is similar to that of (i) and, therefore, is omitted.   □

Lemma 4.4.5 *Let $k \in \mathbb{N}$ be fixed and let $A_0, \ldots, A_k$ be self-adjoint operators affiliated with $\mathcal{M}$. Let $k < r < q < \infty$ and let $\varphi : \mathbb{R}^{k+1} \to \mathbb{C}$ be such that the operator $T_{\varphi}^{A_0,\ldots,A_k} \in \mathcal{B}_k((L^q + L^r)(\mathcal{M}, \tau)^{\times k}, (L^{\frac{q}{k}} + L^{\frac{r}{k}})(\mathcal{M}, \tau))$. For all $X_1, \ldots, X_k \in (L^q + L^r)(\mathcal{M}, \tau)$, $X \in \mathcal{M}$ and for every $A\eta\mathcal{M}_{sa}$ the following assertions hold.*

(i) *Setting*

$$\psi_0(x_0, \ldots, x_{k+1}) := \varphi(x_0, x_2, \ldots, x_{k+1}),$$

we have that

$$T_{\psi_0}^{A_0,A,A_1,\ldots,A_k} \in \mathcal{B}_{k+1}(\mathcal{M} \times (L^q + L^r)(\mathcal{M}, \tau)^{\times k}, (L^{\frac{q}{k}} + L^{\frac{r}{k}})(\mathcal{M}, \tau))$$

and

$$T_{\psi_0}^{A_0,A,A_1,\ldots,A_k}(X, X_1, \ldots, X_k) = T_{\varphi}^{A_0,\ldots,A_k}(XX_1, X_2, \ldots, X_k). \qquad (4.4.3)$$

(ii) *Setting*

$$\psi_1(x_0, x_1, \ldots, x_{k+1}) := \varphi(x_1, \ldots, x_{k+1}),$$

we have that

$$T_{\psi_1}^{A,A_0,\ldots,A_k} \in \mathcal{B}_{k+1}(\mathcal{M} \times (L^q + L^r)(\mathcal{M}, \tau)^{\times k}, (L^{\frac{q}{k}} + L^{\frac{r}{k}})(\mathcal{M}, \tau))$$

and

$$T_{\psi_1}^{A,A_0,\ldots,A_k}(X, X_1, \ldots, X_k) = X \cdot T_{\varphi}^{A_0,\ldots,A_k}(X_1, \ldots, X_k).$$

Proof (i) Let $k = 1$. Fix self-adjoint operators A_0, A_1 affiliated with $\mathcal{M}$ and $1 < r < q < \infty$. Let $\varphi : \mathbb{R}^2 \to \mathbb{C}$ be such that $T_\varphi^{A_0,A_1} \in \mathcal{B}((L^q + L^r)(\mathcal{M}, \tau), (L^q + L^r)(\mathcal{M}, \tau))$. In this case

$$\psi_0(x_0, x_1, x_2) = \varphi(x_0, x_2).$$

Take $X_1 \in (L^q + L^r)(\mathcal{M}, \tau)$, $X \in \mathcal{M}$ and a self-adjoint operator A affiliated with $\mathcal{M}$. Expanding the partial sums of $T_\varphi^{A_0,A_1}(XX_1)$, we obtain

$$S_{\varphi,m}(XX_1) = \sum_{l_0,l_1 \in \mathbb{Z}} \varphi\left(\frac{l_0}{m}, \frac{l_1}{m}\right) E_{A_0}\left(\left[\frac{l_0}{m}, \frac{l_0 + 1}{m}\right)\right) XX_1 E_{A_1}\left(\left[\frac{l_1}{m}, \frac{l_1 + 1}{m}\right)\right)$$

$$= \sum_{l_0,l_1 \in \mathbb{Z}} \varphi\left(\frac{l_0}{m}, \frac{l_1}{m}\right) E_{A_0}\left(\left[\frac{l_0}{m}, \frac{l_0 + 1}{m}\right)\right) X$$

$$\times \left(\sum_{l \in \mathbb{Z}} E_A\left(\left[\frac{l}{m}, \frac{l + 1}{m}\right)\right)\right) X_1 E_{A_1}\left(\left[\frac{l_1}{m}, \frac{l_1 + 1}{m}\right)\right)$$

$$= \sum_{l_0,l,l_1 \in \mathbb{Z}} \psi_0\left(\frac{l_0}{m}, \frac{l}{m}, \frac{l_1}{m}\right) E_{A_0}\left(\left[\frac{l_0}{m}, \frac{l_0 + 1}{m}\right)\right)$$

$$\times X E_A\left(\left[\frac{l}{m}, \frac{l + 1}{m}\right)\right) X_1 E_{A_1}\left(\left[\frac{l_1}{m}, \frac{l_1 + 1}{m}\right)\right)$$

$$= S_{\psi_0,m}(X, X_1), \quad m \in \mathbb{N}.$$

Taking the limit in $(L^q + L^r)(\mathcal{M}, \tau)$, by Definition 4.4.3, we obtain (4.4.3).

Since $\{S_{\varphi,m}\}_{m=1}^\infty$ is a uniformly bounded sequence of linear operators in $\mathcal{B}((L^q + L^r)(\mathcal{M}, \tau), (L^q + L^r)(\mathcal{M}, \tau))$, it follows that

$$\|S_{\psi_0,m}(X, X_1)\|_{L^q + L^r} = \|S_{\varphi,m}(XX_1)\|_{L^q + L^r}$$

$$\leqslant \text{const}\, \|XX_1\|_{L^q + L^r} \leqslant \text{const}\, \|X\|_\infty \|X_1\|_{L^q + L^r}, \quad m \in \mathbb{N}.$$

Thus, $\{S_{\psi_0,m}\}_{m=1}^\infty$ is a uniformly bounded sequence of operators in $\mathcal{B}_2(\mathcal{M} \times (L^q + L^r)(\mathcal{M}, \tau), (L^q + L^r)(\mathcal{M}, \tau))$, and so its limit $T_{\psi_0}^{A_0,A,A_1}$ also belongs to $\mathcal{B}_2(\mathcal{M} \times (L^q + L^r)(\mathcal{M}, \tau), (L^q + L^r)(\mathcal{M}, \tau))$.

The proof for an arbitrary $k > 1$ follows the proof in the case $k = 1$ verbatim.

The proof of (ii) can be derived similarly. $\square$

Norm Estimates

The following bound is a consequence of [24, Lemma 4.6].

Theorem 4.4.6 *Let $n \in \mathbb{N}$, $A_1, \ldots, A_{n+1}$ be self-adjoint operators affiliated with $\mathcal{M}$, $\varphi \in \mathfrak{A}_n$, and $\mathcal{I}$ be a symmetrically normed ideal of $\mathcal{M}$ with property (F). Then,* *there exists $c_I > 0$ such that the multiple operator integral given by Definition 4.3.1 satisfies*

$$\left\| T_\varphi^{A_1,\ldots,A_{n+1}}(X_1,\ldots,X_n) \right\|_I \leqslant c_I \|\varphi\|_{\mathfrak{A}_n} \|X_1\|_I \ldots \|X_n\|_I$$

for all $X_j \in I$, $j = 1,\ldots,n$.

Below we inherit the notation of Theorem 4.3.10 and state the result of Theorem [163, Theorem 5.3] in the full generality.

Theorem 4.4.7 *Let $n \in \mathbb{N}$, $\wp \in \mathcal{P}_n$, $h \in C_b(\mathbb{R})$, and let $\phi_{n,h,\wp}$ be the polynomial integral momentum defined by (4.3.14). Let $1 < p, p_j < \infty$, $j = 1,\ldots,n$, be such that $\frac{1}{p} = \frac{1}{p_1} + \cdots + \frac{1}{p_n}$. Let $A_j \eta \mathcal{M}$, $j = 0,\ldots,n$, and let $T_{\phi_{n,h,\wp}}^{A_0,\ldots,A_n}$ be the transformation given by Definition 4.4.1. Then,*

$$T_{\phi_{n,h,\wp}}^{A_0,\ldots,A_n} \in \mathcal{B}_n(L^{p_1}(\mathcal{M},\tau) \times \cdots \times L^{p_n}(\mathcal{M},\tau), L^p(\mathcal{M},\tau))$$

and there exists $c_{\wp,p_1,\ldots,p_n} > 0$ such that

$$\left\| T_{\phi_{n,h,\wp}}^{A_0,\ldots,A_n} \right\| \leqslant c_{\wp,p_1,\ldots,p_n} \|h\|_\infty.$$

Perturbation Formula

Denote the polynomial integral momentum $\varphi_{n,h,1}$ defined in (4.3.14) by $\varphi_{n,h}$, that is,

$$\varphi_{n,h}(\lambda_0,\ldots,\lambda_n) = \int_{S_n} h\left(\sum_{j=0}^n s_j \lambda_j \right) d\sigma_n. \tag{4.4.4}$$

The following algebraic property extends [161, Theorem 11] to the case of τ-measurable operators; it is proved in [170, Theorem 28].

Theorem 4.4.8 *Let $k \in \mathbb{N}$, $k + 1 < r < q < \infty$. Let $A, B \in (L^q + L^r)(\mathcal{M},\tau)$ and $A_1,\ldots,A_k$ be self-adjoint operators. Let $h \in C_b(\mathbb{R})$ be such that $h' \in C_b(\mathbb{R})$. Then*

$$T_{\varphi_{k,h}}^{A,A_1,\ldots,A_k}(X_1,\ldots,X_k) - T_{\varphi_{k,h}}^{B,A_1,\ldots,A_k}(X_1,\ldots,X_k)$$

$$= T_{\varphi_{k+1,h'}}^{A,B,A_1,\ldots,A_k}(A - B, X_1,\ldots,X_k)$$

holds for all $X_1,\ldots,X_k \in (L^q + L^r)(\mathcal{M},\tau)$, where $\varphi_{k,h}, \varphi_{k+1,h'}$ are given by (4.4.4). If, in addition, f is such that $h = f^{(k)}$, then

$$T_{f^{[k]}}^{A,A_1,\ldots,A_k}(X_1,\ldots,X_k) - T_{f^{[k]}}^{B,A_1,\ldots,A_k}(X_1,\ldots,X_k)$$

$$= T_{f^{[k+1]}}^{A,B,A_1,\ldots,A_k}(A - B, X_1,\ldots,X_k).$$

Proof In the proof we will frequently use [51, Proposition 2.5], implying that if a sequence of projections $\{P_n\}_{n=1}^{\infty} \subset \mathcal{M}$ decreases to 0 in the strong operator topology, $\mathcal{M}$ is atomless, and $X \in L^q(\mathcal{M}, \tau) + L^r(\mathcal{M}, \tau)$, then $\|XP_n\|_{L^q + L^r} \to 0$. By considering $\mathcal{N} \otimes L^{\infty}(0, 1)$ instead of $\mathcal{N}$, we can assume without loss of generality that $\mathcal{N}$ is atomless.

By Theorem 4.4.7 and Lemma 2.6.2,

$$T_{\varphi_{k,h}}^{A,A_1,\ldots,A_k}, T_{\varphi_{k,h}}^{B,A_1,\ldots,A_k} \in \mathcal{B}_k((L^q + L^r)(\mathcal{M}, \tau)^{\times k}, (L^{\frac{q}{k}} + L^{\frac{r}{k}})(\mathcal{M}, \tau)) \qquad (4.4.5)$$

and

$$T_{\varphi_{k+1,h'}}^{A,B,A_1,\ldots,A_k} \in \mathcal{B}_{k+1}((L^q + L^r)(\mathcal{M}, \tau)^{\times(k+1)}, (L^{\frac{q}{k+1}} + L^{\frac{r}{k+1}})(\mathcal{M}, \tau)). \qquad (4.4.6)$$

Denote

$$\psi_0(x_0, \ldots, x_{k+1}) := x_0 \, \varphi_{k+1,h'}(x_0, \ldots, x_{k+1}),$$

$$\psi_1(x_0, \ldots, x_{k+1}) := x_1 \, \varphi_{k+1,h'}(x_0, \ldots, x_{k+1}).$$

Let Q be a projection such that $\tau(Q) < \infty$. Since $A, B \in (L^q + L^r)(\mathcal{M}, \tau)$, it follows that $AQ, QB, AQ - QB \in (L^q + L^r)(\mathcal{M}, \tau)$. Let $A^{(m)} = AE_A([-m, m])$ and $B^{(m)} = BE_B([-m, m])$. By multilinearity of the multiple operator integral, property (4.4.1), and Lemma 4.4.4,

$$T_{\varphi_{k+1,h'}}^{A^{(m)}, B^{(m)}, A_1, \ldots, A_k}((A^{(m)}Q - QB^{(m)}), X_1, \ldots, X_k)$$

$$= T_{\varphi_{k+1,h'}}^{A^{(m)}, B^{(m)}, A_1, \ldots, A_k}(A^{(m)}Q, X_1, \ldots, X_k)$$

$$\quad - T_{\varphi_{k+1,h'}}^{A^{(m)}, B^{(m)}, A_1, \ldots, A_k}(QB^{(m)}, X_1, \ldots, X_k)$$

$$= T_{\psi_0}^{A^{(m)}, B^{(m)}, A_1, \ldots, A_k}(Q, X_1, \ldots, X_k) - T_{\psi_1}^{A^{(m)}, B^{(m)}, A_1, \ldots, A_k}(Q, X_1, \ldots, X_k)$$

$$= T_{\psi_0 - \psi_1}^{A^{(m)}, B^{(m)}, A_1, \ldots, A_k}(Q, X_1, \ldots, X_k). \qquad (4.4.7)$$

By Proposition 4.3.9,

$$\varphi_{k+1,h'}(x_0, \ldots, x_{k+1}) = f^{[k+1]}(x_0, \ldots, x_{k+1})$$

$$= (f^{[k]})^{[1]}(x_0, \ldots, x_{k+1}) = \varphi_{k,h}^{[1]}(x_0, \ldots, x_{k+1}).$$

Hence,

$$(\psi_0 - \psi_1)(x_0, x_1, \ldots, x_{k+1}) = \varphi_{k,h}(x_0, x_2, \ldots, x_{k+1}) - \varphi_{k,h}(x_1, x_2, \ldots, x_{k+1}).$$

Therefore, by Lemma 4.4.5,

$$T_{\psi_0 - \psi_1}^{A^{(m)}, B^{(m)}, A_1, \ldots, A_k}(Q, X_1, \ldots, X_k)$$

$$= T_{\varphi_{k,h}}^{A^{(m)}, A_1, \ldots, A_k}(QX_1, \ldots, X_k) - QT_{\varphi_{k,h}}^{B^{(m)}, A_1, \ldots, A_k}(X_1, \ldots, X_k). \qquad (4.4.8)$$

Combining (4.4.7) and (4.4.8) implies

$$T_{\varphi_{k+1,h'}}^{A^{(m)}, B^{(m)}, A_1, \ldots, A_k}((A^{(m)}Q - QB^{(m)}), X_1, \ldots, X_k)$$

$$= T_{\varphi_{k,h}}^{A^{(m)}, A_1, \ldots, A_k}(QX_1, \ldots, X_k) - QT_{\varphi_{k,h}}^{B^{(m)}, A_1, \ldots, A_k}(X_1, \ldots, X_k). \qquad (4.4.9)$$

By Definition 4.4.1,

$$T_{\varphi_{k+1,h'}}^{A^{(m)}, B^{(m)}, A_1, \ldots, A_k}((A^{(m)}Q - QB^{(m)}), X_1, \ldots, X_k)$$

$$= T_{\varphi_{k+1,h'}}^{A, B, A_1, \ldots, A_k}(E_A([-m,m])(A^{(m)}Q - QB^{(m)})E_B([-m,m]), X_1, \ldots, X_k).$$

Clearly,

$$E_A([-m,m])(A^{(m)}Q - QB^{(m)})E_B([-m,m])$$

$$- E_A([-m,m])(AQ - QB)E_B([-m,m])$$

$$\to AQ - QB$$

in $(L^q + L^r)(\mathcal{M}, \tau)$ as $m \to \infty$. Hence, by (4.4.6),

$$T_{\varphi_{k+1,h'}}^{A^{(m)}, B^{(m)}, A_1, \ldots, A_k}((A^{(m)}Q - QB^{(m)}), X_1, \ldots, X_k)$$

$$\to T_{\varphi_{k+1,h'}}^{A, B, A_1, \ldots, A_k}((AQ - QB), X_1, \ldots, X_k) \qquad (4.4.10)$$

in $(L^{\frac{q}{k+1}} + L^{\frac{r}{k+1}})(\mathcal{M}, \tau)$ as $m \to \infty$. We also have

$$T_{\varphi_{k,h}}^{A^{(m)}, A_1, \ldots, A_k}(QX_1, \ldots, X_k) = E_A([-m,m])T_{\varphi_{k,h}}^{A, A_1, \ldots, A_k}(QX_1, \ldots, X_k).$$

Hence, by (4.4.5),

$$T_{\varphi_{k,h}}^{A^{(m)}, A_1, \ldots, A_k}(QX_1, \ldots, X_k) \to T_{\varphi_{k,h}}^{A, A_1, \ldots, A_k}(QX_1, \ldots, X_k) \qquad (4.4.11)$$

in $(L^{\frac{q}{k}} + L^{\frac{r}{k}})(\mathcal{M}, \tau)$ as $m \to \infty$. Similarly,

$$QT_{\varphi_{k,h}}^{B^{(m)}, A_1, \ldots, A_k}(X_1, \ldots, X_k) \to QT_{\varphi_{k,h}}^{B, A_1, \ldots, A_k}(X_1, \ldots, X_k) \qquad (4.4.12)$$

in $(L^{\frac{q}{k}} + L^{\frac{r}{k}})(\mathcal{M}, \tau)$ as $m \to \infty$. Combining (4.4.10)–(4.4.12) and (4.4.9) gives

$$T^{A,B,A_1,\dots,A_k}_{\varphi_{k+1,h'}}((AQ - QB), X_1, \dots, X_k)$$

$$= T^{A,A_1,\dots,A_k}_{\varphi_{k,h}}(QX_1, \dots, X_k) - QT^{B,A_1,\dots,A_k}_{\varphi_{k,h}}(X_1, \dots, X_k). \qquad (4.4.13)$$

Since τ is a semifinite trace, it follows that there exists a sequence $\{Q_n\}_{n\in\mathbb{N}}$ of projections satisfying $Q_n \uparrow I$ and $\tau(Q_n) < \infty$. By (4.4.13),

$$T^{A,B,A_1,\dots,A_k}_{\varphi_{k+1,h'}}((AQ_n - Q_nB), X_1, \dots, X_k)$$

$$= T^{A,A_1,\dots,A_k}_{\varphi_{k,h}}(Q_nX_1, \dots, X_k) - Q_nT^{B,A_1,\dots,A_k}_{\varphi_{k,h}}(X_1, \dots, X_k), \quad n \in \mathbb{N}.$$
$$(4.4.14)$$

Since $AQ_n - Q_nB \to A - B$ as $n \to \infty$, in $(L^q + L^r)(\mathcal{M}, \tau)$, it follows that

$$T^{A,B,A_1,\dots,A_k}_{\varphi_{k+1,h'}}((AQ_n - Q_nB), X_1, \dots, X_k) \to T^{A,B,A_1,\dots,A_k}_{\varphi_{k+1,h'}}((A - B), X_1, \dots, X_k)$$

in $(L^{\frac{q}{k+1}} + L^{\frac{r}{k+1}})(\mathcal{M}, \tau)$, and so also with respect to the measure topology. Similarly, since $Q_nX_1 \to X_1$ in $(L^q + L^r)(\mathcal{M}, \tau)$ and

$$Q_nT^{B,A_1,\dots,A_k}_{\varphi_{k,h}}(X_1, \dots, X_k) \to T^{B,A_1,\dots,A_k}_{\varphi_{k,h}}(X_1, \dots, X_k)$$

in $(L^{\frac{q}{k}} + L^{\frac{r}{k}})(\mathcal{M}, \tau)$, it follows that

$$T^{A,A_1,\dots,A_k}_{\varphi_{k,h}}(Q_nX_1, \dots, X_k) - Q_nT^{B,A_1,\dots,A_k}_{\varphi_{k,h}}(X_1, \dots, X_k)$$

$$\to T^{A,A_1,\dots,A_k}_{\varphi_{k,h}}(X_1, \dots, X_k) - T^{B,A_1,\dots,A_k}_{\varphi_{k,h}}(X_1, \dots, X_k)$$

in $(L^{\frac{q}{k}} + L^{\frac{r}{k}})(\mathcal{M}, \tau)$, and so also with respect to the measure topology. Taking the limit in (4.4.14) with respect to the measure topology we complete the proof of the theorem. $\square$

Hölder-Type Estimates

The main result of this section is the Hölder-type estimate given in Theorem 4.4.11, which is established in [170, Theorem 34] and extends [161, Corollary 13 and Theorem 14]. In the proof of Theorem 4.4.11, the interpolation argument of [161, Theorem 14] is replaced with a different technique based on the Calderón-type operator $P_{q,r}$ defined in (4.4.15) below. Such operators provide a useful technical tool in many questions of interpolation theory (see, e.g., [28, Chapter 3, Section 5]).

For $1 < q, r < \infty$ and $X \in S(\mathcal{M}, \tau)$, consider the operator $P_{q,r} : S(\mathcal{M}, \tau) \to S(0, \infty)$, given by

$$(P_{q,r}(X))(t) := \left(\frac{1}{t} \int_0^t \mu_s^r(X) \, ds\right)^{1/r} + \left(\frac{1}{t} \int_t^\infty \mu_s^q(X) \, ds\right)^{1/q}, \quad t > 0.$$

(4.4.15)

Observe that if $X \in (L^r + L^q)(\mathcal{M}, \tau)$, $r < q$, then the value $P_{q,r}(X)(t)$ is a finite number for all $t > 0$. Observe also that there is $t > 0$ such that $P_{q,r}(X)(t) = 0$ if and only if $X = 0$.

Lemma 4.4.9 *If* $1 < r < q < \infty$, *then for* $p \in (r, q)$ *there exists a constant* $c(p, q, r) > 0$ *such that*

$$\|P_{q,r}(X)\|_{L^{p,\infty}} \leqslant c(p, q, r)\|X\|_{L^{p,\infty}}, \quad X \in L^{p,\infty}(\mathcal{M}, \tau).$$

Proof Let $X \in L^{p,\infty}(\mathcal{M}, \tau)$, $t > 0$. Then,

$$(P_{q,r}(X))(t) = \left(\frac{1}{t} \int_0^t \mu_s^r(X) ds\right)^{1/r} + \left(\frac{1}{t} \int_t^\infty \mu_s^q(X) ds\right)^{1/q}$$

$$\leqslant \sup_{s>0} s^{\frac{1}{p}} \mu_s(X) \cdot \left(\left(\frac{1}{t} \int_0^t s^{-\frac{r}{p}} ds\right)^{1/r} + \left(\frac{1}{t} \int_t^\infty s^{-\frac{q}{p}} ds\right)^{1/q}\right)$$

$$- \|X\|_{L^{p,\infty}}' \left(\left(\frac{1}{t} \int_0^t s^{-\frac{r}{p}} ds\right)^{1/r} + \left(\frac{1}{t} \int_t^\infty s^{-\frac{q}{p}} ds\right)^{1/q}\right)$$

$$= t^{-\frac{1}{p}} \|X\|_{L^{p,\infty}}' \left(\left(\frac{p}{p-r}\right)^{\frac{1}{r}} + \left(\frac{p}{q-p}\right)^{\frac{1}{q}}\right).$$

Appealing to (2.6.1) completes the proof.   □

The main technical tool in the proof of Theorem 4.4.11 is the estimate obtained in the next result due to [170, Theorem 33].

Let σ_u, $u \in (0, \infty)$, denote the dilation operator

$$\sigma_u(f)(s) = f\left(\frac{s}{u}\right).$$

We note that σ_u is a bounded linear operator on the Banach space $L^{p,\infty}(0, \infty)$ for $1 < p < \infty$ with the norm

$$\|\sigma_u\|_{L^{p,\infty} \to L^{p,\infty}} = u^{\frac{1}{p}}.$$

(4.4.16)

Theorem 4.4.10 *Let* $k \in \mathbb{N}$ *and let* $k + 1 < r < q < \infty$. *Then, there exists a constant* $C(k, q, r) > 0$ *such that for a pair of self-adjoint operators* $A, B \in (L^r + L^q)(\mathcal{M}, \tau)$, *for all self-adjoint operators* $A_1, \ldots, A_k$ *affiliated with* $\mathcal{M}$ *and* *every compactly supported function* $h \in \Lambda_\alpha$, $\alpha \in [0, 1]$,

$$\mu\big(T_{\varphi_{k,h}}^{A,A_1,\ldots,A_k}(X_1,\ldots,X_k) - T_{\varphi_{k,h}}^{B,A_1,\ldots,A_k}(X_1,\ldots,X_k)\big)$$

$$\leqslant C(k,q,r)\|h\|_{\Lambda_\alpha}\sigma_3(P_{r,q}(A-B))^\alpha \prod_{j=1}^{k} \sigma_3(P_{r,q}(X_j))$$

holds for all $X_1,\ldots,X_k \in (L^r + L^q)(\mathcal{M},\tau)$, *where* $\varphi_{k,h}$ *is given by* (4.4.4).

The proof of Theorem 4.4.10 involves Theorems 4.4.7, 4.4.8, and Lemma 2.6.2. We refer the reader for details to [170, Theorem 33].

The following Hölder-type estimate for multiple operator integrals is due to [170, Theorem 35].

Theorem 4.4.11 *Let* $k \in \mathbb{N}$ *and let* $k + 1 < p < \infty$. *Then, there exists a constant* $C(k, p) > 0$ *such that for every compactly supported function* $h \in \Lambda_\alpha$, $\alpha \in [0, 1]$, *and all self-adjoint operators* $A, B \in L^{p,\infty}(\mathcal{M},\tau)$

$$\Big\|T_{\varphi_{k,h}}^{A,A_1,\ldots,A_k}(X_1,\ldots,X_k) - T_{\varphi_{k,h}}^{B,A_1,\ldots,A_k}(X_1,\ldots,X_k)\Big\|_{L^{\frac{p}{k+\alpha},\infty}}$$

$$\leqslant C(k,p)\cdot\|h\|_{\Lambda_\alpha}\, \|A-B\|_{L^{p,\infty}}^\alpha \Big(\prod_{j=1}^{k}\|X_j\|_{L^{p,\infty}}\Big)$$

holds for all $X_1,\ldots,X_k \in L^{p,\infty}(\mathcal{M},\tau)$, *where* $\varphi_{k,h}$ *is given by* (4.4.4).

Proof Let $k \in \mathbb{N}$, $k + 1 < p < \infty$ and $\alpha \in [0, 1]$ be fixed. Observe that $A, B, X_1,\ldots,X_k \in (L^r + L^q)(\mathcal{M},\tau)$ for $r = \frac{1}{2}(p + k + 1)$ and $q = 2p$.

Denote for brevity

$$D := T_{\varphi_{k,h}}^{A,A_1,\ldots,A_k}(X_1,\ldots,X_k) - T_{\varphi_{k,h}}^{B,A_1,\ldots,A_k}(X_1,\ldots,X_k).$$

By Theorem 4.4.10,

$$\mu(D) \leqslant C(k,q,r)\|h\|_{\Lambda_\alpha}\cdot \sigma_3(P_{r,q}(A-B))^\alpha \cdot \prod_{j=1}^{k}\sigma_3(P_{r,q}(X_j)). \tag{4.4.17}$$

Taking the norm $\|\cdot\|_{L^{\frac{p}{k+\alpha},\infty}}$ on both sides of (4.4.17) and applying Lemma 2.6.1(ii) gives

$$\|D\|_{L^{\frac{p}{k+\alpha},\infty}} \leqslant, C(k,q,r)\, C'(k,p)\, \|h\|_{\Lambda_\alpha}\big\|\sigma_3(P_{r,q}(A-B))^\alpha\big\|_{L^{\frac{p}{\alpha},\infty}}$$

$$\cdot \prod_{j=1}^{k}\big\|\sigma_3(P_{r,q}(X_j))\big\|_{L^{p,\infty}},$$

where $C'(k, p) = \max\limits_{0 \leqslant \alpha \leqslant 1} c_2\left(k + 1, \frac{p}{k+\alpha}\right)$ and $c_2(m, q) = \frac{q}{q-1} m^{\frac{1}{q}}$. Recalling (4.4.16), we infer that

$$\|D\|_{L^{\frac{p}{k+\alpha}},\infty} \leqslant 3^{\frac{k+\alpha}{p}} C(k, q, r) C'(k, p) \|h\|_{\Lambda_\alpha} \left\|(P_{r,q}(A - B))^\alpha\right\|_{L^{\frac{p}{\alpha}},\infty}$$

$$\cdot \prod_{j=1}^{k} \left\|P_{r,q}(X_j)\right\|_{L^{p,\infty}}$$

$$\leqslant 3^{\frac{k+1}{p}} C(k, q, r) C'(k, p) \|h\|_{\Lambda_\alpha} \left\|P_{r,q}(A - B)\right\|_{L^{p,\infty}}^{\alpha}$$

$$\cdot \prod_{j=1}^{k} \left\|P_{r,q}(X_j)\right\|_{L^{p,\infty}}.$$

Hence, by Lemma 4.4.9,

$$\|D\|_{L^{\frac{p}{k+\alpha}},\infty} \leqslant 3^{\frac{k+1}{p}} C(k, q, r) C'(k, p) c(p, q, r)^{k+\alpha} \|h\|_{\Lambda_\alpha} \|A - B\|_{L^{p,\infty}}^{\alpha}$$

$$\cdot \prod_{j=1}^{k} \|X_j\|_{L^{p,\infty}}$$

$$\leqslant C''(k, p) \|h\|_{\Lambda_\alpha} \|A - B\|_{L^{p,\infty}}^{\alpha} \cdot \prod_{j=1}^{k} \|X_j\|_{L^{p,\infty}},$$

where $C''(k, p) = 3^{\frac{k+1}{p}} C(k, q, r) C'(k, p) \max\limits_{0 \leqslant \alpha \leqslant 1} c(p, q, r)^{k+\alpha}$. $\square$

The following assertion due to [170, Theorem 35] generalizes Theorem 4.4.11 and also extends the result of [4, Theorem 5.8] (see Sect. 5.2) for Schatten class perturbations to a multilinear setting. Its proof is completely analogous to the proof of Theorem 4.4.11 and, therefore, it is omitted.

Theorem 4.4.12 *Let $\alpha \in [0, 1]$, $k \in \mathbb{N}$, and let $1 < p, p_j < \infty$, $j = 0, \ldots, k$ be such that $\frac{1}{p} = \frac{\alpha}{p_0} + \frac{1}{p_1} + \cdots + \frac{1}{p_k}$. Then, there are constants $c(k, p_1, \ldots, p_k), \tilde{c}(k, p_1, \ldots, p_k) > 0$ such that for every compactly supported function $h \in \Lambda_\alpha$ and*

(i) *for all $A = A^*, B = B^* \in L^{p_0}(\mathcal{M}, \tau)$, $X_j \in L^{p_j}(\mathcal{M}, \tau)$, $j = 1, \ldots, k$,*

$$\left\|T_{\varphi_{k,h}}^{A, A_1, \ldots, A_k}(X_1, \ldots, X_k) - T_{\varphi_{k,h}}^{B, A_1, \ldots, A_k}(X_1, \ldots, X_k)\right\|_{L^p}$$

$$\leqslant c(k, p_1, \ldots, p_k) \cdot \|h\|_{\Lambda_\alpha} \|A - B\|_{L^{p_0}}^{\alpha} \left(\prod_{j=1}^{k} \|X_j\|_{L^{p_j}}\right);$$

(ii) *for all* $A = A^*, B = B^* \in L^{p_0,\infty}(\mathcal{M}, \tau)$, $X_j \in L^{p_j,\infty}(\mathcal{M}, \tau)$, $j = 1, \ldots, k$,

$$\left\| T_{\varphi_{k,h}}^{A,A_1,\ldots,A_k}(X_1, \ldots, X_k) - T_{\varphi_{k,h}}^{B,A_1,\ldots,A_k}(X_1, \ldots, X_k) \right\|_{L^{p,\infty}}$$

$$\leqslant \tilde{c}(k, p_1, \ldots, p_k) \cdot \|h\|_{\Lambda_\alpha} \|A - B\|_{L^{p_0,\infty}}^{\alpha} \left(\prod_{j=1}^{k} \|X_j\|_{L^{p_j,\infty}} \right),$$

where $\varphi_{k,h}$ *is given by* (4.4.4).

The following strong technical result is established in [170, Theorem 36] by utilizing a two-dimensional induction and multiple operator integration techniques, including Theorems 4.4.8 and 4.4.11.

Theorem 4.4.13 *Let* $m \in \mathbb{N}$, $m \geqslant 2$ *and* $p \in (m, m + 1]$. *Then, there exists a constant* $c(p) > 0$ *such that for every compactly supported function* g *on* $\mathbb{R}$ *satisfying* $g^{(j)}(0) = 0$, $j = 0, \ldots, m - 1$, *and* $g^{(m-1)} \in \Lambda_{p-m}$,

$$\left\| T_{g^{[k]}}^{A_0,\ldots,A_k} \right\|_{(L^{p,\infty})^{\times k} \to L^{\frac{p}{p-1},\infty}} \leqslant c(p) \cdot \|g^{(m-1)}\|_{\Lambda_{p-m}} \left(\sum_{j=0}^{k} \|A_j\|_{L^{p,\infty}} \right)^{p-k-1}$$

holds for all self-adjoint elements $A_0 \ldots, A_k \in L^{p,\infty}(\mathcal{M}, \tau)$.

Chapter 5
Applications

In this chapter we discuss various results of operator theory, functional analysis, mathematical physics, and noncommutative geometry that rely on methods of multiple operator integration.

5.1 Operator Lipschitz Functions

The study of operator Lipschitz functions emerged from the now answered question of M. G. Krein [115]. A detailed exposition on operator Lipschitzness with respect to the operator and trace class norms is given in [9]; for a brief summary see [149]. In this section we briefly discuss major results on operator Lipschitzness with respect to the operator and Schatten norms.

Initially the operator Lipschitz functions were introduced in the case of self-adjoint operators, but the definition naturally generalizes to the nonself-adjoint case.

Definition 5.1.1 Let I be an interval in $\mathbb{R}$, $f : I \to \mathbb{C}$ a continuous function, and $1 \leqslant p \leqslant \infty$. We say that f is operator Lipschitz on I with respect to the norm $\| \cdot \|_p$ if there exists a constant $c_{f,p} > 0$ such that

$$\| f(A) - f(B) \|_p \leqslant c_{f,p} \| A - B \|_p$$

for all $A, B \in \mathcal{D}_{sa}$ with $\sigma(A) \cup \sigma(B) \subset I$ and every separable Hilbert space $\mathcal{H}$. We will briefly call the operator Lipschitz functions with respect to the S^p-norm "operator S^p-Lipschitz functions" and operator Lipschitz functions with respect to the $\mathcal{B}(\mathcal{H})$-norm "operator Lipschitz functions".

© Springer Nature Switzerland AG 2019
A. Skripka, A. Tomskova, *Multilinear Operator Integrals*,
Lecture Notes in Mathematics 2250, https://doi.org/10.1007/978-3-030-32406-3_5

5.1.1 Commutator and Lipschitz Estimates in $\mathcal{S}^2$

Proposition 5.1.2 *Let $A, B \in \mathcal{B}_{sa}(\mathcal{H})$ and $\sigma(A) \cup \sigma(B) \subseteq [a, b]$. If $f \in \mathrm{Lip}[a, b]$, then*

$$\|f(A)X - Xf(B)\|_2 \leqslant \|f\|_{\mathrm{Lip}[a,b]} \|AX - XB\|_2, \quad X \in \mathcal{S}^2.$$

Proof The result is an immediate consequence of Theorem 3.3.6 and the property

$$\|T_{f^{[1]}}^{A,B} : \mathcal{S}^2 \to \mathcal{S}^2\| = \|f^{[1]}\|_\infty = \|f\|_{\mathrm{Lip}[a,b]}$$

following from (3.2.2) and (3.2.3). $\square$

The following result is an immediate consequence of Proposition 3.2.2 and Theorem 3.3.7. Nonetheless, we demonstrate an independent proof of it based on a similar result for finite matrices (see [83, Theorem 4.1]).

Theorem 5.1.3 *For every $A, B \in \mathcal{D}_{sa}$ such that $A - B \in \mathcal{S}^2$ and $f \in \mathrm{Lip}(\mathbb{R})$, the estimate*

$$\|f(A) - f(B)\|_2 \leqslant \|f\|_{\mathrm{Lip}(\mathbb{R})} \|A - B\|_2 \tag{5.1.1}$$

holds, that is, f is operator Lipschitz on $\mathbb{R}$ with respect to the $\mathcal{S}^2$-norm.

Proof (Proof in the Case $A, B \in \mathcal{B}_{sa}(\mathcal{H})$) Let $\{\xi_j\}_{j=1}^\infty$ be an orthonormal basis in $\mathcal{H}$ and P_N the orthogonal projection onto the linear span of $\{\xi_j\}_{j=1}^N$, $N \in \mathbb{N}$. Firstly we observe that

$$\|P_N(A - B)P_N\|_2^2 = \sum_{j,k=1}^\infty \left|\langle P_N(A - B)P_N\xi_j, \xi_k\rangle\right|^2$$

$$= \sum_{j,k=1}^\infty \left|\langle (A - B)P_N\xi_j, P_N\xi_k\rangle\right|^2$$

$$= \sum_{j,k=1}^N \left|\langle (A - B)\xi_j, \xi_k\rangle\right|^2 \leqslant \|A - B\|_2^2.$$

Hence, from the estimate (3.1.11) obtained for finite matrices, we have

$$\|f(P_N A P_N) - f(P_N B P_N)\|_2 \leqslant \|f\|_{\mathrm{Lip}(\mathbb{R})} \|P_N A P_N - P_N B P_N\|_2$$

$$\leqslant \|f\|_{\mathrm{Lip}(\mathbb{R})} \|A - B\|_2.$$

Thus, for any $n \in \mathbb{N}$, it follows that

$$\sum_{j,k=1}^{n} \left| \langle (f(P_N A P_N) - f(P_N B P_N)) \xi_j, \xi_k \rangle \right|^2 \leqslant \|f\|_{\mathrm{Lip}(\mathbb{R})}^2 \|A - B\|_2^2.$$

Since $P_N A P_N \rightarrow A$ and $P_N B P_N \rightarrow B$ in the strong operator topology as $N \rightarrow \infty$, by [171, Theorem VIII.20(b)] we infer that $f(P_N A P_N) \rightarrow f(A)$ and $f(P_N B P_N) \rightarrow f(B)$ in the strong operator topology. Thus,

$$\sum_{j,k=1}^{n} \left| \langle (f(A) - f(B)) \xi_j, \xi_k \rangle \right|^2 \leqslant \|f\|_{\mathrm{Lip}(\mathbb{R})}^2 \|A - B\|_2^2, \quad \text{for any } n \in \mathbb{N}.$$

Taking $n \rightarrow \infty$, we obtain $f(A) - f(B) \in \mathcal{S}^2$ and the inequality (5.1.1).   □

5.1.2   Commutator and Lipschitz Estimates in $\mathcal{S}^p$ and $\mathcal{B}(\mathcal{H})$

The class of operator Lipschitz functions in $\mathcal{S}^p$, $1 < p < \infty$, coincides with the set of scalar Lipschitz functions, while the set of operator Lipschitz functions in $\mathcal{S}^1$ and $\mathcal{B}(\mathcal{H})$ is smaller. The details are discussed below.

The next result is a consequence of [41, Theorem 8.2] and the bound for the double operator integral (3.3.6). In the particular case of $A, B \in \mathcal{B}_{sa}(\mathcal{H})$ with $\sigma(A) \cup \sigma(B) \subseteq [a, b]$, the proof of this result goes along the lines of the proof of Theorem 3.3.6 and applies properties of the double operator integral discussed in Sect. 3.3.5.

Theorem 5.1.4 *Let $A, B \in \mathcal{D}_{sa}$ be such that $A - B \in \mathcal{B}(\mathcal{H})$ and let $X \in \mathcal{B}(\mathcal{H})$. If $f \in \mathrm{Lip}(\mathbb{R})$ is such that $f^{[1]} \in \mathfrak{A}_1$, then*

$$f(A)X - Xf(B) = T_{f^{[1]}}^{A,B}(AX - XB) \tag{5.1.2}$$

and

$$\|f(A)X - Xf(B)\| \leqslant \|f^{[1]}\|_{\mathfrak{A}_1} \|AX - XB\|. \tag{5.1.3}$$

We note that applying (5.1.2) to $X = I$ recovers the representation (3.3.9) and applying (5.1.3) to $X = I$ gives the estimate below.

Theorem 5.1.5 *Let $1 \leqslant p \leqslant \infty$ and $A, B \in \mathcal{D}_{sa}$ be such that $A - B \in \mathcal{S}^p$ (or $A - B \in \mathcal{B}(\mathcal{H})$ if $p = \infty$). If $f \in \mathrm{Lip}(\mathbb{R})$ such that $f^{[1]} \in \mathfrak{A}_1$, then*

$$\|f(A) - f(B)\|_p \leqslant \|f^{[1]}\|_{\mathfrak{A}_1} \|A - B\|_p. \tag{5.1.4}$$

We have the following necessary and sufficient conditions for operator Lipschitz-ness described in terms of harmonic analysis.

Theorem 5.1.6

(i) *Every function* $f \in B^1_{\infty 1}(\mathbb{R})$ *is operator Lipschitz on* $\mathbb{R}$ *with respect to the operator and Schatten norms.*

(ii) *If* f *is operator Lipschitz on* $\mathbb{R}$ *with respect to the operator and trace class norms, then* $f \in B^1_{11}(\mathbb{R})_{loc}$.

Proof The property (i), which was established in [143], is an immediate consequence of Theorems 3.3.14 and 3.3.8.

The property (ii) is established in [141].   □

Remark 5.1.7 If f is as in Theorem 3.3.8, then

$$\| f \|_{\mathrm{Lip}(\mathbb{R})} \leqslant \| f^{[1]} \|_{\mathfrak{A}_1}$$

by the straightforward estimate

$$| f^{[1]}(\lambda, \mu) | \leqslant \int_\Omega |a_1(\lambda, \omega)| \cdot |a_2(\lambda, \omega)| \, d|\nu|(\omega)$$

$$\leqslant \int_\Omega \| a_1(\cdot, \omega) \|_\infty \| a_2(\cdot, \omega) \|_\infty \, d|\nu|(\omega).$$

Therefore, the constant in the estimate (5.1.4) is worse than in the case $p = 2$ (established in Theorem 5.1.3).

The restriction $f^{[1]} \in \mathfrak{A}_1$ in Theorem 5.1.5 is removed in [159, Theorem 1] in the case $p \neq 1$, implying that every Lipschitz function is operator Lipschitz with respect to the Schatten S^p-norm, $p > 1$.

Theorem 5.1.8 *Let* $1 < p < \infty$ *and* I *be a (bounded or unbounded) interval in* $\mathbb{R}$. *A continuous function* $f : I \to \mathbb{C}$ *is operator Lipschitz with respect to the* S^p-*norm on* I *if and only if* $f \in \mathrm{Lip}(I)$. *Moreover, there is* $c_p > 0$ *such that*

$$\| f(A) - f(B) \|_p \leqslant c_p \| f \|_{\mathrm{Lip}(I)} \| A - B \|_p$$

for all $A, B \in \mathcal{D}_{sa}$ *with* $\sigma(A) \cup \sigma(B) \subseteq I$ *and all* $f \in \mathrm{Lip}(I)$.

Proof (Proof Outline) If f is operator Lipschitz in S^p, then it is scalar Lipschitz. This immediately follows from considering operators on a one dimensional Hilbert space.

Assume now that $f \in \mathrm{Lip}(I)$. Adjusting the estimate for the double operator integral in Theorem 3.3.4 to the interval I and combining it with the result of [156, Theorem 5.3] implies

$$\| f(U)V - Vf(U) \|_p \leqslant c_p \| UV - VU \|_p \tag{5.1.5}$$

for every self-adjoint operator U and every bounded operator V. Applying (5.1.5) to

$$U = \begin{pmatrix} A & 0 \\ 0 & B \end{pmatrix} \text{ and } V = \begin{pmatrix} 0 & I \\ I & 0 \end{pmatrix}$$

implies the result. □

Remark 5.1.9 It was proved in [62] that the space of all operator Lipschitz functions with respect to the Schatten norm $\| \cdot \|_p$, $1 < p < \infty$, contains nondifferentiable functions, for example, $f(t) = |t|$. The fact that the absolute value function $f(t) = |t|$ is not operator Lipschitz on $\mathbb{R}$ with respect to $\| \cdot \|_1$ and $\| \cdot \|$ was proved earlier in [62, 99]. More generally, it follows from the results in [93] and from [104, Corollary 3.7] that all functions that are operator Lipschitz with respect to the operator norm $\| \cdot \|$ are differentiable. There are also continuously differentiable Lipschitz functions that are not operator Lipschitz with respect to $\| \cdot \|_1$ [80, 131, 208]. These functions are also not operator Lipschitz with respect to $\| \cdot \|$ because operator Lipschitzness with respect to the operator norm is equivalent to the operator Lipschitzness with respect to the trace class norm [9, Theorem 3.6.5]. The converse problem whether there exist operator Lipschitz functions with respect to $\| \cdot \|$ that are not continuously differentiable was posed in [213] and found affirmative answers in [104, 109] as detailed below.

One of the ways to construct functions that are not operator Lipschitz is to find a sequence of finite matrices of increasing dimension so that the respective Lipschitz bounds grow logarithmically with the dimension and then take appropriate direct sums of such matrices. This strategy is at the heart of the counterexample for the function $f(t) = |t|$ outlined in [62].

The following finite-dimensional result for the trace class norm is obtained in [62, Theore 13] and for the operator norm in [8, Remark after Theorem 11.4] with involvement of [62, Lemma 15].

Theorem 5.1.10

(i) *For every* $d \in \mathbb{N}$ *there exist* $A_d, B_d \in \mathcal{B}_{sa}(\ell_{2d}^2)$ *such that* $A_d \neq B_d$ *and*

$$\| |A_d| - |B_d| \|_1 \geqslant \text{const} \cdot \log d \cdot \|A_d - B_d\|_1.$$

(ii) *For every* $d \in \mathbb{N}$ *there exist* $A_d, B_d \in \mathcal{B}_{sa}(\ell_d^2)$ *such that* $A_d \neq B_d$ *and*

$$\| |A_d| - |B_d| \| \geqslant \text{const} \cdot (1 + \log d) \cdot \|A_d - B_d\|.$$

Below we provide another finite-dimensional construction that is suitable to build higher order counterexamples for Taylor remainders with bounded operators in Sect. 5.4. The result is derived in [169, Theorem 5.2] from [54, Theorem 7], while the latter is based on results of [208].

Theorem 5.1.11 *For every* $d \in \mathbb{N}$, $d \geqslant 2$, *there exist non-zero self-adjoint operators* A_d, $B_d \in \mathcal{B}(\ell_{2d}^2)$ *such that* $\sigma(A_d + B_d) = \sigma(A_d) \subset [-e^{-1}, e^{-1}]$, $0 \in \sigma(A_d)$ *has multiplicity* 2, *every* $\lambda \in \sigma(A_d) \setminus \{0\}$ *has multiplicity* 1, *and*

$$\|h(A_d + B_d) - h(A_d)\| \geqslant \mathrm{const}\, (\log d)^{\frac{1}{2}} \|B_d\|,$$

where

$$h = \begin{cases} |x|\big(\log |\log |x| - 1|\big)^{-\frac{1}{2}}, & x \in [-e^{-1}, e^{-1}] \setminus \{0\} \\ 0, & x = 0 \end{cases} \tag{5.1.6}$$

is a function in $C^1(\mathbb{R})$.

The following conditions for f to be operator Lipschitz on $[a, b]$ are due to [109, Corollary 4.6].

Theorem 5.1.12 *Suppose that* $f \in C[a, b]$ *and there are* $x_n \searrow a$, $x_0 = b$, *such that* f *is operator Lipschitz on each segment* $I_n = [x_n, x_{n-1}]$, *that is,*

$$\|f(A) - f(B)\| \leqslant c_n \|A - B\| \tag{5.1.7}$$

for all A, $B \in \mathcal{B}_{sa}(\mathcal{H})$ *such that* $\sigma(A) \cup \sigma(B) \subseteq I_n$ *and some* $c_n > 0$. *Then,* f *is operator Lipschitz on* $[a, b]$ *if and only if*

$$\sum_{n=1}^{\infty} \left(\frac{f_n}{x_n - x_{n-1}}\right)^2 < \infty, \quad \text{where } f_n = \sup\{|f(x) - f(a)| : x \in [a, x_{n-1}]\},$$

and $\sup\limits_{n \in \mathbb{N}} c_n < \infty$, *where* c_n *satisfy* (5.1.7).

Based on Theorem 5.1.12 one can construct a large variety of operator Lipschitz functions that are not continuously differentiable. The first example of such function is given in [104, Theorem 3.8]:

$$f(t) = \begin{cases} t^2 \sin\left(\frac{1}{t}\right) & \text{if } t \neq 0 \\ 0 & \text{if } t = 0. \end{cases}$$

The following result is obtained in [109, Corollary 5.2].

Theorem 5.1.13 *Let* φ *be an infinitely many times differentiable, nonnegative function on* $\mathbb{R}$ *such that*

$$\mathrm{supp}(\varphi) = [-1, 1], \quad \max_{t \in \mathbb{R}} \varphi(t) = 1, \quad \varphi'(-1/2) = 1, \quad \varphi'(1/2) = -1.$$

Let $\{\sigma_n\}_{n=1}^{\infty}$ in $\mathbb{R}_+$ be such that $\sigma_n \searrow 0$, $\sigma_1 < \frac{1}{4}$ and $\sum_{n=1}^{\infty} \sigma_n^2 < \infty$. Set $d_n = \frac{3}{2^{n+1}}$, $a_n = \frac{\sigma_n}{2^n}$ and $\varphi_n(t) = a_n \varphi\left(\frac{t-d_n}{a_n}\right)$. Then, the function $g(t) := \sum_{n=1}^{\infty} \varphi_n(t)$ is infinitely many times differentiable on $\mathbb{R} \setminus \{0\}$, differentiable but not continuously at $t = 0$, and operator Lipschitz on $\operatorname{supp}(g)$.

Although not every perturbation in $\mathcal{S}^1$ produces an increment of an operator function in $\mathcal{S}^1$, this increment belongs to the larger ideal $\mathcal{S}^{1,\infty}$, provided the respective scalar function is Lipschitz. The following analog of Theorem 5.1.8 is a consequence of Theorem 3.3.17.

Theorem 5.1.14 *Let $f \in \operatorname{Lip}(\mathbb{R})$. Then, there exists an absolute constant $c > 0$ such that*

$$\|f(A) - f(B)\|_{1,\infty} \leqslant c \, \|f\|_{\operatorname{Lip}(\mathbb{R})} \, \|A - B\|_1$$

for all $A, B \in \mathcal{D}_{sa}$.

5.1.3 Commutator and Lipschitz Estimates: Nonself-adjoint Case

Operator Lipschitzness of functions of unitary operators is completely analogous to the one of self-adjoint operators. The following analog of Theorem 5.1.8 for unitaries is established in [21, Theorem 2].

Theorem 5.1.15 *Let A, B be unitaries, $1 < p < \infty$. Then, there is $c_p > 0$ such that*

$$\|f(A) - f(B)\|_p \leqslant c_p \, \|f\|_{\operatorname{Lip}(\mathbb{T})} \, \|A - B\|_p, \tag{5.1.8}$$

where

$$\|f\|_{\operatorname{Lip}(\mathbb{T})} = \sup_{x \neq y \in \mathbb{T}} \frac{|f(x) - f(y)|}{|x - y|}.$$

An analog of Theorem 5.1.6 for functions of unitary operators is established in [141].

Theorem 5.1.16

(i) *Every function $f \in B_{\infty 1}^1(\mathbb{T})$ is operator Lipschitz on $\mathbb{T}$ with respect to the operator and Schatten norms.*

(ii) *If f is operator Lipschitz on $\mathbb{T}$ with respect to the operator and trace class norms, then $f \in B_{11}^1(\mathbb{T})$.*

An example of a function in $C^1(\mathbb{T})$ that is not operator Lipschitz in $\mathcal{S}^1$ on $\mathbb{T}$ is given in [141]. Similarly to a self-adjoint case, such example can also be constructed based on dimension dependent bounds for matrix functions. The following dimension dependent bound is derived in [54, Theorem 8].

Theorem 5.1.17 *For every integer $d \geqslant 3$, there exist unitary operators H_d, $K_d \in \mathcal{B}(\ell_{2d+1}^2)$ such that $\sigma(H_d) = \sigma(K_d)$, $1 \in \sigma(H_d)$, and*

$$\|u(K_d) - u(H_d)\| \geqslant \text{const} \, (\log d)^{\frac{1}{2}} \|K_d - H_d\|,$$

where u is given by

$$u(e^{i\theta}) := \tilde{h}(\theta) \tag{5.1.9}$$

and $\tilde{h}$ is a 2π-periodic function in $C^1(\mathbb{R}) \cap C^n(\mathbb{R} \setminus \{0\})$ extending the function h defined by (5.1.6).

Below we state results on Lipschitzness of functions of normal operators. The first one is proved in [110, Corollary 6.1].

Theorem 5.1.18 *Let $1 < p < \infty$ and I be a compact set in $\mathbb{C}$. A continuous function f on $\mathbb{C}$ is operator Lipschitz with respect to the S^p-norm on the set*

$$\mathcal{B}_{norm}(\mathcal{H})(I) := \{A \in \mathcal{B}(\mathcal{H}) : \ A \text{ is normal}, \ \sigma(A) \subseteq I\} \tag{5.1.10}$$

if and only if $f \in \text{Lip}(I)$. Moreover, there is $c_p > 0$ such that

$$\|f(A) - f(B)\|_p \leqslant c_p \, \|f\|_{\text{Lip}(I)} \, \|A - B\|_p \tag{5.1.11}$$

for all A, $B \in \mathcal{B}_{norm}(\mathcal{H})(I)$ and all $f \in \text{Lip}(I)$, where

$$\|f\|_{\text{Lip}(I)} = \sup_{x \neq y \in I} \frac{|f(x) - f(y)|}{\|x - y\|_1}.$$

A bound similar to (5.1.11) is also obtained for the commutator $f(A)X - Xf(A)$ in [110, Corollary 6.1].

The following operator norm bound for a quasicommutator $f(A)X - Xf(B)$ is due to [13, Theorem 10.3].

Theorem 5.1.19 *Let A, $B \in \mathcal{B}(\mathcal{H})$ be normal operators and $X \in \mathcal{B}(\mathcal{H})$. Then,*

$$\|f(A)X - Xf(B)\| \leqslant c \, \|f\|_{B_{\infty 1}^1(\mathbb{R}^2)} \max\{\|AX - XB\|, \|A^*X - XB^*\|\}.$$

The appropriate adjustment of the latter result also holds for unbounded normal operators A, B.

We note that the aforementioned results are based on theory of double operator integrals, each one employing different aspects of this theory. To achieve their result, the authors of [13] studied operator Lipschitzness of functions of two variables.

The investigation of the operator Lipschitzness of functions of two variables in the operator and Schatten-von Neumann norms was continued in [11, 14].

Operator Lipschitzness has also been studied for contractions and dissipative operators. The following analog of (5.1.8) for contractions is proved in [110, Theorem 6.4].

Theorem 5.1.20 *Let A, B be contractions, $1 < p < \infty$, and $f \in \mathcal{A}(\mathbb{D})$. Then,*

$$\|f(A) - f(B)\|_p \leqslant c_p \|f\|_{\mathrm{Lip}(\mathbb{D})} \|A - B\|_p,$$

where

$$\|f\|_{\mathrm{Lip}(\mathbb{D})} = \sup_{x \neq y \in \mathbb{D}} \frac{|f(x) - f(y)|}{\|x - y\|_1}.$$

It is established in [107, Theorem 3.4] that every function $f \in \mathcal{A}(\mathbb{D})$ that is operator Lipschitz on $\mathbb{T}$ (for all pairs of unitaries) with respect to the operator norm is also operator Lipschitz on $\mathbb{D}$ (for all pairs of contractions) and, moreover,

$$\|f(A)X - Xf(B)\| \leqslant \|f\|_{\mathrm{OL}(\mathbb{T})} \|AX - XB\|$$

where A, B are contractions, $X \in \mathcal{B}(\mathcal{H})$, and

$$\|f\|_{\mathrm{OL}(\mathbb{T})} = \sup \left\{ \frac{\|f(A) - f(B)\|}{\|B - A\|} : A \neq B \text{ are unitaries} \right\}.$$

It is proved in [12, Theorem 5.3] that

$$\|f(A)X - Xf(B)\| \leqslant \|f\|_{\mathrm{OL}(\mathbb{C}_+)} \|AX - XB\|$$

where $X \in \mathcal{B}(\mathcal{H})$, A, B are maximal dissipative operators with bounded quasicommutator $AX - XB$, and

$$\|f\|_{\mathrm{OL}(\mathbb{C}_+)} = \sup \left\{ \frac{\|f(A) - f(B)\|}{\|B - A\|} : A \neq B \text{ are normal with } \sigma(A) \cup \sigma(B) \subset \bar{\mathbb{C}}_+ \right\}.$$

5.1.4 Lipschitz Type Estimates in Noncommutative L^p-Spaces

The results of Theorems 5.1.8 and 3.3.17 extend to the setting of noncommutative L^p spaces, as it is done in [159, Theorem 1] and [49, Theorem 5.3].

Theorem 5.1.21 *Let $(\mathcal{M}, \tau)$ be a semifinite von Neumann algebra and $A \eta \mathcal{M}_{sa}$, $B \eta \mathcal{M}_{sa}$. Let $f \in \mathrm{Lip}(\mathbb{R})$ and $1 < p < \infty$. Then, there exists $c_p > 0$ such that*

$$\|f(A) - f(B)\|_{L^p(\mathcal{M}, \tau)} \leqslant c_p \|f\|_{\mathrm{Lip}(\mathbb{R})} \|A - B\|_{L^p(\mathcal{M}, \tau)}.$$

Theorem 5.1.22 *Let $(\mathcal{M}, \tau)$ be a semifinite von Neumann algebra and $A\eta\mathcal{M}_{sa}$, $B\eta\mathcal{M}_{sa}$. If $f \in \mathrm{Lip}(\mathbb{R})$, then there exists $c > 0$ such that*

$$\|f(A) - f(B)\|_{L^{1,\infty}(\mathcal{M},\tau)} \leqslant c \, \|f\|_{\mathrm{Lip}(\mathbb{R})} \, \|A - B\|_{L^1(\mathcal{M},\tau)}.$$

5.1.5   Lipschitz Type Estimates in Banach Spaces

Let $\mathcal{X}, \mathcal{Y}$ be Banach spaces. We are interested in the Lipschitz type estimates

$$\|f(B) - f(A)\|_{\mathcal{B}(\mathcal{X})} \leqslant c_{A,B,f} \, \|B - A\|_{\mathcal{B}(\mathcal{X})} \qquad (5.1.12)$$

for $A, B \in \mathcal{B}(\mathcal{X})$ and, more generally, commutator estimates

$$\|f(B)X - Xf(A)\|_{\mathcal{B}(\mathcal{X},\mathcal{Y})} \leqslant c_{A,B,f} \, \|BX - XA\|_{\mathcal{B}(\mathcal{X},\mathcal{Y})} \qquad (5.1.13)$$

for $A \in \mathcal{B}(\mathcal{X})$, $B \in \mathcal{B}(\mathcal{Y})$, $X \in \mathcal{B}(\mathcal{X}, \mathcal{Y})$. As it was mentioned above, this problem is well-known in the special case when $\mathcal{X} = \mathcal{Y}$ is a separable Hilbert space, such as ℓ^2, and A and B are normal operators on $\mathcal{X}$. Here we present such estimates in the Banach space setting, and specifically for $\mathcal{X} = \ell^p$ and $\mathcal{Y} = \ell^q$ with $p, q \in [1, \infty]$. For all relevant definitions see Sect. 2.10.

The following result is due to [173, Theorem 4.6].

Theorem 5.1.23 *Let $\mathcal{X}, \mathcal{Y}$ be separable Banach spaces such that either $\mathcal{X}$ or $\mathcal{Y}$ has a bounded approximation property and let $\mathcal{I}$ be a Banach ideal in $\mathcal{B}(\mathcal{X}, \mathcal{Y})$ with a strong convex compactness property. If $A, B \in \mathcal{B}_s(\mathcal{X})$ and $f^{[1]} \in \mathfrak{A}_1$, then*

$$\|f(B)X - Xf(A)\|_{\mathcal{I}} \leqslant 16 \operatorname{spec}(A) \operatorname{spec}(B) \, \|f^{[1]}\|_{\mathfrak{A}_1} \|BX - XA\|_{\mathcal{I}}$$

and, in particular,

$$\|f(B) - f(A)\|_{\mathcal{I}} \leqslant 16 \operatorname{spec}(A) \operatorname{spec}(B) \, \|f^{[1]}\|_{\mathfrak{A}_1} \|B - A\|_{\mathcal{I}}.$$

It is immediate from the definition of a scalar type operator that every normal operator on $\mathcal{H}$ is of scalar type, which extends Theorem 3.3.8 for $p = \infty$ to the Banach space setting.

If A and B are diagonalizable operators, then the class of functions in the above theorem is extended in [173, Theorem 7.3]. Given Banach spaces $\mathcal{X}, \mathcal{Y}$ and $1 \leqslant p < \infty$, let Π_p denote the ideal in $\mathcal{B}(\mathcal{X}, \mathcal{Y})$ consisting of all $S : \mathcal{X} \to \mathcal{Y}$ such that for every $n \in \mathbb{N}$ and every collection $\{x_j\}_{j=1}^n \subset \mathcal{X}$,

$$\left(\sum_{j=1}^n \|S(x_j)\|_{\mathcal{Y}}^p \right)^{1/p} \leqslant C \sup_{\|x^*\|_{\mathcal{X}^*} \leqslant 1} \left(\sum_{j=1}^n |\langle x^*, x_j \rangle|^p \right)^{1/p}$$

The infimum of C as above gives a norm on Π_p, which we denote by π_p. The ideal (Π_p, π_p) is called the ideal of p-summing operators from X to $\mathcal{Y}$; it is the Banach ideal by [67, Propositions 2.3, 2.4, 2.6].

Theorem 5.1.24 *Let* $1 < p < \infty$ *and* $\frac{1}{p} + \frac{1}{p^*} = 1$. *Let* $A \in \mathcal{B}(\ell^{p^*})$ *(respectively,* $A \in \mathcal{B}(c_0)$) *and* $B \in \mathcal{B}(\ell^p)$ *(respectively,* $B \in \mathcal{B}(\ell^1)$) *be diagonalizable operators. Let* $(\mathcal{I}, \|\cdot\|_{\mathcal{I}})$ *be the ideal of* p-summing operators from ℓ^{p^*} to ℓ^p *(respectively, from* c_0 *to* ℓ^1). *Then, every* $f \in \mathrm{Lip}(\mathbb{C})$ *satisfies* (5.1.13) *with* $c_{A,B,f} = c_{A,B} \|f\|_{\mathrm{Lip}(\mathbb{C})}$.

Analogous results for different pairs of spaces (ℓ^p, ℓ^q) are derived in [173, Theorems 6.8 and 6.9].

Theorem 5.1.25 *Let* $1 \leqslant p < q < \infty$. *Let* $A \in \mathcal{B}(\ell^p)$ *and* $B \in \mathcal{B}(\ell^q)$ *(respectively,* $B \in \mathcal{B}(c_0)$) *be diagonalizable operators with real spectra. Then,* (5.1.13) *holds with* $\mathcal{B}(X, \mathcal{Y}) = \mathcal{B}(\ell^p, \ell^q)$ *(respectively,* $\mathcal{B}(X, \mathcal{Y}) = \mathcal{B}(\ell^p, c_0)$) *and* $f(t) = |t|$, *where* $c_{A,B,f} = c_{A,B} \|f\|_{\mathrm{Lip}(\mathbb{R})}$.

Theorem 5.1.26 *Let* $1 \leqslant p < q < \infty$. *Let* $A \in \mathcal{B}(\ell^1)$ *and* $B \in \mathcal{B}(\ell^q)$ *(respectively,* $B \in \mathcal{B}(c_0)$) *be diagonalizable operators. Then,* (5.1.13) *holds with* $\mathcal{B}(X, \mathcal{Y}) = \mathcal{B}(\ell^1, \ell^q)$ *(respectively,* $\mathcal{B}(X, \mathcal{Y}) = \mathcal{B}(\ell^p, c_0)$) *for every Lipschitz function* f, *where* $c_{A,B,f} = c_{A,B} \|f\|_{\mathrm{Lip}(\mathbb{C})}$. *In addition,* (5.1.12) *holds with* $X = \ell^1$ *and* $X = c_0$.

The proofs in [173] rely on the theory of Schur multipliers on the space $\mathcal{B}(\ell^p, \ell^q)$ developed by G. Bennett [26, 27] and double operator integrals discussed in Sect. 3.6. Commutator estimates for $f(t) = |t|$ and different Banach ideals in $\mathcal{B}(\mathcal{H})$ were also studied in [62, 70], where the proofs are based on Macaev's celebrated theorem (see [85]) or on the UMD-property of the reflexive Schatten-von Neumann ideals. However, the spaces $\mathcal{B}(X, \mathcal{Y})$ are not UMD-spaces and, therefore, the techniques used in [62, 70] do not apply to them.

5.1.6 Operator $\mathcal{I}$-Lipschitz Functions

Operator $\mathcal{I}$-Lipschitz and Commutator $\mathcal{I}$-Bounded Functions

Let $\mathcal{I}$ be a symmetrically normed (s. n.) ideal of $\mathcal{B}(\mathcal{H})$ equipped with the norm $\|\cdot\|_{\mathcal{I}}$. Denote by $\mathcal{I}_{norm}$ the set of all normal operators in $\mathcal{I}$ and by I a compact subset of $\mathbb{C}$. Set

$$\mathcal{I}_{norm}(I) := \{A \in \mathcal{I}_{norm} : \sigma(A) \subseteq I\}.$$

Definition 5.1.27

(i) $f \in C(\mathbb{C})$ is called an $\mathcal{I}$-Lipschitz function on $I \subset \mathbb{C}$ if there is $D > 0$ such that $f(A) - f(B) \in \mathcal{I}$ and

$$\|f(A) - f(B)\|_{\mathcal{I}} \leqslant D \|A - B\|_{\mathcal{I}}, \quad A, B \in \mathcal{I}_{norm}(I). \tag{5.1.14}$$

(ii) f is a commutator $\mathcal{I}$-bounded function on $I \subset \mathbb{C}$ if there is $D > 0$ such that, for all $A \in \mathcal{I}_{norm}(I)$ and $X \in \mathcal{B}(\mathcal{H})$, we have $f(A)X - Xf(A) \in \mathcal{I}$ and

$$\|f(A)X - Xf(A)\|_{\mathcal{I}} \leqslant D \, \|AX - XA\|_{\mathcal{I}} \, . \tag{5.1.15}$$

We note that if $\mathcal{I} = \mathcal{B}(\mathcal{H})$, then f is an operator Lipschitz function. The spaces of all $\mathcal{I}$-Lipschitz and commutator $\mathcal{I}$-bounded functions on the interval I are denoted $\mathcal{I}$-Lip(I) and $\mathcal{I}$-CB(I), respectively.

The result below is due to [104, Theorem 3.5].

Theorem 5.1.28 *Let $f \in C(\mathbb{C})$, $I \subset \mathbb{C}$ and let $\mathcal{I}$ be an s. n. ideal. The following properties are equivalent.*

(i) f is an $\mathcal{I}$-Lipschitz function on I.
(ii) (5.1.14) holds for all $A, B \in \mathfrak{F}_{norm}(I)$.
(iii) (5.1.15) holds for all $A \in \mathcal{I}_{norm}(I)$, all $X \in \mathcal{B}_{sa}(\mathcal{H})$.
(iv) (5.1.15) holds for all $A \in \mathfrak{F}_{norm}(I)$, all $X \in \mathfrak{F}_{sa}$.

Condition (5.1.15) is equivalent to the following stronger condition (see [104, Proposition 4.1]): f is commutator $\mathcal{I}$-bounded on I if and only if there is $D > 0$ such that, for all $A, B \in \mathcal{I}_{norm}(I)$ and $X \in \mathcal{B}(\mathcal{H})$, we have $f(A)X - Xf(B) \in \mathcal{I}$ and

$$\|f(A)X - Xf(B)\|_{\mathcal{I}} \leqslant D \, \|AX - XB\|_{\mathcal{I}} \, . \tag{5.1.16}$$

For all s. n. ideals, including $\mathcal{I} = \mathcal{B}(\mathcal{H})$, [104, Corollaries 3.6 and 5.4] and (5.1.16) yield the following result.

Corollary 5.1.29

(i) $\mathcal{I}$-CB$(I) \subseteq \mathcal{I}$-Lip$(I)$ for $I \subset \mathbb{C}$ and $\mathcal{I}$-CB$(I) = \mathcal{I}$-Lip(I) for $I \subset \mathbb{R}$.
(ii) $\mathcal{B}(\mathcal{H})$-CB$(I) = \mathcal{S}^{\infty}$-CB$(I) = \mathcal{S}^{1}$-CB$(I)$ for all $I \subset \mathbb{C}$.

For $\mathcal{I} = \mathcal{S}^p$, $1 < p < \infty$, the above results were noticed in [62]. By Corollary 5.1.29(i), the condition that $f \in \mathcal{I}$-CB(I) is stronger than the condition that $f \in \mathcal{I}$-Lip(I) for $I \subset \mathbb{C}$. If $I \subset \mathbb{R}$ these conditions are equivalent. The possibility to reduce the study of $\mathcal{I}$-Lipschitz functions to the study of commutator $\mathcal{I}$-bounded functions is important since it enables us to use interpolation theory techniques.

Definition 5.1.30

(i) A compact set $I \subset \mathbb{C}$ is called $\mathcal{I}$-Fuglede if $\mathcal{I}$-CB$(I) = \mathcal{I}$-Lip(I).
(ii) A s. n. ideal $\mathcal{I}$ is called a Fuglede ideal if all compacts I in $\mathbb{C}$ are $\mathcal{I}$-Fuglede.

Proposition 5.1.31 ([104, Proposition 4.5]) *A compact $I \subset \mathbb{C}$ is $\mathcal{I}$-Fuglede if and only if the function $h(z) = \bar{z}$ is commutator $\mathcal{I}$-bounded on I, that is, there is $D > 0$ such that*

$$\left\|A^*X - XA^*\right\|_{\mathcal{I}} \leqslant D \, \|AX - XA\|_{\mathcal{I}} \, , \quad A \in \mathcal{I}_{norm}(I) \text{ and } X \in \mathcal{B}(\mathcal{H}).$$

The following sufficient conditions for an ideal to be Fuglede are obtained in [105, Corollary 3.8]; for the definition of Boyd indices see, for instance, [16].

Theorem 5.1.32 *Let I be a separable s. n. ideal and (p_I, q_I) be its Boyd indices. If $1 < p_I, q_I < \infty$, then I is a Fuglede ideal.*

The theorem above, in particular, implies that all S^p, $1 < p < \infty$, are Fuglede ideals (see also [1, 182, 211]).

The following results are obtained in [105, Theorem 4.3 and Corollary 4.6].

Theorem 5.1.33 *Let I be a separable s. n. ideal.*

(i) *If $f \in I\text{-CB}(I)$ for $I \subset \mathbb{C}$, then there exists $D > 0$ such that (5.1.16) holds for all $A, B \in \mathcal{B}_{norm}(\mathcal{H})(I)$ satisfying $AX - XB \in I$ for all $X \in \mathcal{B}(\mathcal{H})$, where $\mathcal{B}_{norm}(\mathcal{H})(I)$ is defined in (5.1.10).*

(ii) *If $f \in I\text{-Lip}(I)$ and $I \subset \mathbb{C}$ is I-Fuglede compact, then (5.1.16) holds for all $A, B \in \mathcal{B}_{norm}(\mathcal{H})(I)$ satisfying $AX - XB \in I$ for all $X \in \mathcal{B}(\mathcal{H})$. In particular, $A - B \in I$ implies $f(A) - f(B) \in I$ and $\|f(A) - f(B)\|_I \leqslant D \|A - B\|_I$.*

Apart from separable ideals, Theorem 5.1.33 holds for a large variety of other s. n. ideals (see [105, Theorem 4.5 and Corollary 4.6]).

It is established in [105, Corollary 3.8] that the ideals $S^1, S^\infty, \mathcal{B}(\mathcal{H})$ are not Fuglede by testing the properties of $f(z) = \bar{z}$ and using the result of [92] that there are $A \in \mathcal{B}_{norm}(\mathcal{H})$ and $X \in \mathcal{B}(\mathcal{H})$ such that

$$AX - XA \in S^1, \text{ but } A^*X - XA^* \notin S^1. \tag{5.1.17}$$

It is established in [211] that the operator X in (5.1.17) can be chosen compact. Further, it is proved in [102, Corollary 4.3] that X can be chosen in any S^p for $p > 1$. It is shown in [105, Corollary 5.9] that both A and X can be chosen compact.

Initially, it was thought that all spaces $S^p\text{-Lip}(\mathbb{R})$, $p \in (1, \infty)$, are different. However, it is shown in [159] that they are all the same and coincide with the space $\text{Lip}(\mathbb{R})$. In [110] this result is extended to the spaces $S^p\text{-Lip}(\mathbb{C})$, $p \in (1, \infty)$, which is shown to coincide with the space $\text{Lip}(\mathbb{C})$.

Problem 5.1.34 Let I be a separable s. n. ideal. Does $I\text{-Lip}(\mathbb{C})$ coincide with $\text{Lip}(\mathbb{C})$?

I-Stable and Commutator I-Stable Functions

The notions of I-stable and commutator I-stable functions were introduced in [105]. They are close to the notions of I-Lipschitz and commutator I-bounded functions.

Definition 5.1.35

(i) $f \in C(\mathbb{C})$ is called I-stable on a compact $I \subset \mathbb{C}$ if, for $A, B \in \mathcal{B}_{norm}(\mathcal{H})(I)$, the condition $A - B \in I$ implies $f(A) - f(B) \in I$.

(ii) $f \in C(\mathbb{C})$ is called commutator $\mathcal{I}$-stable on I if $AX - XA \in \mathcal{I}$ for $A \in \mathcal{B}_{norm}(\mathcal{H})(I)$ and all $X \in \mathcal{B}(\mathcal{H})$ implies $A^*X - XA^* \in \mathcal{I}$.

(iii) A compact I is called weakly $\mathcal{I}$-Fuglede if $h(z) = \bar{z}$ is a commutator $\mathcal{I}$-stable function on I.

(iv) A s. n. ideal $\mathcal{I}$ is called weakly Fuglede if all compacts I in $\mathbb{C}$ are weakly $\mathcal{I}$-Fuglede.

In general, the condition of commutator $\mathcal{I}$-stability is stronger than $\mathcal{I}$-stability. In fact, $f \in C(\mathbb{C})$ is $\mathcal{I}$-stable on I if and only if the implication in Definition 5.1.35(ii) holds for all $X = X^* \in \mathcal{B}(\mathcal{H})$ (see [105, Proposition 5.2]). If, however, $\mathcal{I}$ is weakly $\mathcal{I}$-Fuglede, then commutator $\mathcal{I}$-stability coincides with $\mathcal{I}$-stability (see [105, Proposition 5.5]). Clearly, all $I \subset \mathbb{R}$ are weakly $\mathcal{I}$-Fuglede for all ideals $\mathcal{I}$. We also have the following result, which proof is left as an exercise.

Proposition 5.1.36 *Let $\mathcal{I}$ be a separable s. n. ideal.*

(i) If I is a $\mathcal{I}$-Fuglede compact, then I is weakly $\mathcal{I}$-Fuglede.
(ii) If $\mathcal{I}$ is Fuglede, then $\mathcal{I}$ is weakly Fuglede.

Stability and Fuglede properties of the ideals $\mathcal{S}^p$, $1 \leqslant p \leqslant \infty$, and $\mathcal{B}(\mathcal{H})$ are summarized below.

Theorem 5.1.37

(i) Let $1 < p < \infty$. For each compact $I \subset \mathbb{C}$,

$$\mathcal{S}^p\text{-Lip}(I) = \mathcal{S}^p\text{-CB}(I) = \text{Lip}(I).$$

(ii) $\mathcal{S}^\infty$, $\mathcal{B}(\mathcal{H})$ are weakly Fuglede, but not Fuglede ideals. $\mathcal{S}^1$ is not weakly Fuglede.

(iii) For each compact $I \subset \mathbb{C}$, $\mathcal{S}^1\text{-CB}(I) = \mathcal{S}^\infty\text{-CB}(I) = \mathcal{B}(\mathcal{H})\text{-CB}(I) \subsetneqq \text{Lip}(I)$.

(iv) The ideals $\mathcal{S}^1$, $\mathcal{S}^\infty$, $\mathcal{B}(\mathcal{H})$ have the same Fuglede compacts. If I is one of them, then

$$\mathcal{S}^1\text{-Lip}(I) = \mathcal{S}^\infty\text{-Lip}(I) = \mathcal{B}(\mathcal{H})\text{-Lip}(I) = \mathcal{S}^1\text{-CB}(I) = \mathcal{S}^\infty\text{-CB}(I)$$

$$= \mathcal{B}(\mathcal{H})\text{-CB}(I) = \{g \in C(I) \colon g \text{ is commutator } \mathcal{S}^1\text{-stable on } I\}.$$

If $I \subset \mathbb{R}$ then the above spaces coincide with $\{g \in C(I) \colon g \text{ is } \mathcal{S}^1\text{-stable on } I\}$.

$\mathcal{A}$-Lipschitz Functions on Semisimple Hermitian Banach *-Algebras $\mathcal{A}$

Let $\mathcal{A}$ be a semisimple Hermitian Banach *-algebra and $\mathcal{A}_{sa}$ be the set of all self-adjoint elements in $\mathcal{A}$. Denote by $C^*(\mathcal{A})$ the C*-algebra completion of $\mathcal{A}$ with respect to Ptak-Rajkov C*-norm: $\|A\|_r = r_{\mathcal{A}}(A^*A)^{1/2}$ for $A \in \mathcal{A}$, where $r_{\mathcal{A}}(A)$ is the spectral radius of A in $\mathcal{A}$. For all relevant algebraic notions we refer the reader to [60].

Each $f \in C(\mathbb{R})$ acts on $C^*(\mathcal{A})_{sa}$, that is, for each $A \in C^*(\mathcal{A})_{sa}$, there is $f(A) \in C^*(\mathcal{A})$. For $I \subseteq \mathbb{R}$, set

$$\mathcal{A}_{sa}(I) := \{A \in \mathcal{A}_{sa} : \sigma(A) \subseteq I\}.$$

A function $f \in C(\mathbb{R})$ *acts on* $\mathcal{A}_{sa}(I)$ if $f(A) \in \mathcal{A}$ (instead of $f \in C^*(\mathcal{A})$) for all $A \in \mathcal{A}_{sa}(I)$. For example, all S^p, $1 \leqslant p < \infty$, are semisimple Hermitian Banach *-algebras and $C^*(S^p) = S^\infty$. A function $f \in C(\mathbb{R})$ acts on all S^p if and only if $f(0) = 0$ and $\left|\frac{f(t)}{t}\right| < C$ for some $C > 0$ and t in a neighbourhood of 0 (see [108, Theorem 2.2]).

If δ is a weakly closed *-derivation on S^p, $1 \leqslant p < \infty$, its domain $D(\delta)$ is a semisimple Hermitian Banach *-algebra with norm $\|A\|_\delta = \|A\| + \|\delta(A)\|$, $A \in D(\delta)$, and $C^*(D(\delta)) = S^\infty$. If f is Lipschitz on each compact in $\mathbb{R}$ and $f(0) = 0$ then, by [108, Theorem 3.8], f acts on $D(\delta)$. If δ is a weakly closed *-derivation on a C*-algebra $\mathcal{A}$, then $D(\delta)$ is a semisimple Hermitian Banach *-algebra and $C^*(D(\delta)) = \mathcal{A}$. It is shown in [103, Theorem 8.4] that $f \in C(\mathbb{R})$ acts on $D(\delta)$ if and only if f is operator Lipschitz.

Definition 5.1.38 A function $f \in C(\mathbb{R})$ is called $\mathcal{A}$-Lipschitz on $I_0 \subseteq \mathbb{R}$ if it acts on $\mathcal{A}_{sa}(I_0)$ and, for each compact $I \subseteq I_0$, there is $D_I > 0$ such that

$$\|f(A) - f(B)\|_\mathcal{A} \leqslant D_I \|A - B\|_\mathcal{A}$$

for all $A, B \in \mathcal{A}_{sa}(I)$.

Denote by Γ an open subset of $\mathbb{R}$ containing 0 and by $\mathcal{A}$-Lip(Γ) the space of all functions $\mathcal{A}$-Lipschitz on each compact in Γ. The presence of non-trivial $\mathcal{A}$-Lipschitz functions reflects the structure of the algebra $\mathcal{A}$. The following result is due to [106, Corollary 2.5 and Theorem 4.4].

Theorem 5.1.39 *Let $\mathcal{A}$ be a semisimple Hermitian Banach *-algebra.*

(i) *Let $\mathcal{A}$ be unital. If all infinitely differentiable functions on $\mathbb{R}$ act on $\mathcal{A}_{sa}(\mathbb{R})$, then $\mathcal{A}$ is C*-equivalent, that is, it is a unital C*-algebra in some equivalent norm $(C^*(\mathcal{A}) = \mathcal{A})$.*

(ii) *Let $\mathcal{A}$ be unital. If there is a non-linear $\mathcal{A}$-Lipschitz function on some $I = [a, b]$ which extends to a function on $\mathbb{C}$ analytic in a neighbourhood of I in $\mathbb{C}$, then $\mathcal{A}$ is C*-equivalent.*

(iii) *Let $\mathcal{A}$ be not unital. If $f(t) = t^2$ is an $\mathcal{A}$-Lipschitz function on some $0 \in \Gamma \subseteq \mathbb{R}$, then $\mathcal{A}$ is a dense symmetrically normed Jordan ideal of $C^*(\mathcal{A})$, that is, there is $K > 0$ such that $AX + XA \in \mathcal{A}$ and*

$$\|AX + XA\|_\mathcal{A} \leqslant K \|A\| \|X\|_\mathcal{A}$$

for all $A \in C^*(\mathcal{A})$, $X \in \mathcal{A}$.

It is proved in [82] that if C*($\mathcal{A}$) is S^∞ or a properly infinite W*-algebra, then all s. n. Jordan ideals are two-sided ideals of S^∞. Various conditions when $\mathcal{A}$ is a s. n. two-sided ideal of S^∞ are given in [106].

Let $\Pi(\mathcal{A})$ be the set of all irreducible *-representations of $\mathcal{A}$. For $\pi \in \Pi(\mathcal{A})$, let H_π be the representation space. The following result is proved in [106, Theorem 3.3].

Theorem 5.1.40 *Let $\mathcal{A}$ be a C*-algebra and $0 \in \Gamma \subseteq \mathbb{R}$.*

(i) *Let* $\dim H_\pi \leqslant N$ *for some* $N \in \mathbb{N}$ *and all* $\pi \in \Pi(\mathcal{A})$. *Then, the space $\mathcal{A}$-Lip(Γ) coincides with the space of all functions g on Γ such that $g|_I \in$ Lip(I) for all compacts $I \subset \Gamma$.*

(ii) *If N in (i) does not exist, then $\mathcal{A}$-Lip$(\Gamma) = \mathcal{B}(\mathcal{H})$-Lip$(\Gamma)$.*

If $\mathcal{A}$ is a s. n. Jordan ideal of C*($\mathcal{A}$), the space $\mathcal{A}$-Lip(Γ) is described in [106, Theorem 4.20].

Theorem 5.1.41 *Let $\mathcal{A}$ be a s. n. Jordan ideal of a separable C*-algebra C*($\mathcal{A}$). Let $0 \in \Gamma \subseteq \mathbb{R}$.*

(i) *If C*($\mathcal{A}$) is not a CCR-algebra, then $\mathcal{A}$-Lip$(\Gamma) = \mathcal{B}(\mathcal{H})$-Lip$(\Gamma)$.*

(ii) *Let C*($\mathcal{A}$) be a CCR-algebra. Then, for each $\pi \in \Pi(\mathcal{A})$, there is a s. n. ideal J^π of S^∞ such that $J^\pi \neq S^1$, $J^\pi \neq S^\infty$ and $\mathcal{A}$-Lip$(\Gamma) \subseteq J^\pi$-Lip(Γ).*

5.2 Operator Hölder Functions

Operator functions inherit the Hölder property from the respective scalar Hölder functions, which is quite different from the operator Lipschitzness discussed in the previous section. Below we discuss several Hölder-type inequalities for operator functions.

The following results are established in [5, Theorem 4.1]; see also [9].

Theorem 5.2.1 *Let $A, B \in \mathcal{D}_{sa}$ and $f \in \Lambda_\alpha(\mathbb{R})$ for $\alpha \in (0, 1)$. Then, there exists a constant $c > 0$ such that*

$$\|f(A) - f(B)\| \leqslant c(1-\alpha)^{-1} \|f\|_{\Lambda_\alpha(\mathbb{R})} \|A - B\|^\alpha.$$

Operator Hölderness of functions of self-adjoint operators with perturbations in Schatten classes is established in [4, Theorem 5.8].

Theorem 5.2.2 *Let $A, B \in \mathcal{D}_{sa}$ and $f \in \Lambda_\alpha(\mathbb{R})$ for $\alpha \in (0, 1)$, and $1 < p < \infty$. Then, there exists a constant $c > 0$ such that*

$$\|f(A) - f(B)\|_{p/\alpha} \leqslant c \|f\|_{\Lambda_\alpha(\mathbb{R})} \|A - B\|_p^\alpha.$$

The following operator Hölder property in the unitary case is due to [5, Theorem 5.1].

Theorem 5.2.3 *Let A, B be unitary operators and $f \in \Lambda_\alpha(\mathbb{T})$ for $\alpha \in (0, 1)$. Then, there exists a constant $c > 0$ such that*

$$\|f(A) - f(B)\| \leqslant c(1 - \alpha)^{-1} \|f\|_{\Lambda_\alpha(\mathbb{T})} \|A - B\|^\alpha.$$

A result completely analogous to the one of Theorem 5.2.3 for contractions A, B and functions $f \in \Lambda_\alpha(\mathbb{T})$ analytic in the unit disc is derived in [5, Section 6].

The following result is established in [13, Theorem 9.1].

Theorem 5.2.4 *Let A, B be normal operators with the same domain in $\mathcal{H}$, let $f \in \Lambda_\alpha(\mathbb{R}^2)$ for $\alpha \in (0, 1)$, and $1 < p < \infty$. Then, there exists a constant $c_{p,\alpha} > 0$ such that*

$$\|f(A) - f(B)\|_{p/\alpha} \leqslant c_{p,\alpha} \|f\|_{\Lambda_\alpha(\mathbb{R}^2)} \|A - B\|_p^\alpha.$$

Modifications of Theorem 5.2.4 for $p = 1$ and for a quasinormed ideal with upper Boyd index less than 1 are discussed in [13, Section 9].

5.3   Differentiation of Operator Functions

Differentiability is one of natural properties studied in theory of functions, in particular, operator functions. Pioneering results on differentiability of operator functions were obtained in [61], with restrictive assumptions on functions and operators in the infinite-dimensional setting. These results were substantially refined and extended in the series of papers [18, 19, 24, 36, 55, 63, 102, 110, 119, 141, 143, 146, 147, 163, 199] in response to development of perturbation theory and also influenced by the question published in [212].

In this section, we prove the best known results on differentiability of matrix functions and outline the proof of differentiability of operator functions in the infinite-dimensional case. We also outline the proof of best estimates for Schatten norms of operator derivatives obtained in [163].

5.3.1   Differentiation of Matrix Functions

Following the proof of [61, Theorem 1] and supplementing omitted details, we demonstrate below that $t \mapsto f(X(t))$ is differentiable in the operator norm and the respective derivative can be written as a double operator integral (3.1.1).

Proposition 5.3.1 *Let $t \mapsto X(t)$ be a C^1-function with values in $\mathcal{B}_{sa}(\ell_d^2)$. If $f \in C^1(\mathbb{R})$, then the function $t \mapsto f(X(t))$ is differentiable in the operator norm and*

$$\frac{df(X(t))}{dt} = T_{f^{[1]}}^{X(t),X(t)}\left(\frac{dX(t)}{dt}\right). \tag{5.3.1}$$

Proof Let $\{\lambda_j(t)\}_{j=1}^d$ be a complete list of eigenvalues and $\{\xi_j(t)\}_{j=1}^d$ a respective orthonormal basis of eigenvectors of $X(t)$.

Let $f(\lambda) = \lambda^m$, $m \in \mathbb{N}$. Then, differentiating by the product rule gives

$$\frac{df(X(t))}{dt} = \frac{dX^m(t)}{dt} = \sum_{i=0}^{m-1} X^i(t)\frac{dX(t)}{dt}X^{m-i-1}(t).$$

By the spectral theorem, the latter equals

$$\sum_{i=0}^{m-1}\sum_{j=1}^d \lambda_j^i P_{\xi_j} \frac{dX(t)}{dt} \sum_{k=1}^d \lambda_k^{m-i-1} P_{\xi_k}$$

$$= \sum_{j=1}^d \sum_{k=1}^d \left(\sum_{i=0}^{m-1} \lambda_j^i \lambda_k^{m-i-1}\right) P_{\xi_j} \frac{dX(t)}{dt} P_{\xi_k}$$

$$= \begin{cases} \sum_{j=1}^d \sum_{k=1}^d \frac{\lambda_j^m - \lambda_k^m}{\lambda_j - \lambda_k} P_{\xi_j} \frac{dX(t)}{dt} P_{\xi_k}, & \lambda_j \neq \lambda_k, \\ \sum_{j=1}^d \sum_{k=1}^d m\lambda_j^{m-1} P_{\xi_j} \frac{dX(t)}{dt} P_{\xi_k}, & \lambda_j = \lambda_k \end{cases},$$

where we suppressed the parameter t from the notation of the spectral projection and eigenvalues. Applying (3.1.1) completes the proof of (5.3.1) for a monomial. By linearity, (5.3.1) extends from monomials to general polynomials.

Now we extend (5.3.1) to a general $f \in C^1(\mathbb{R})$ via approximations. Let $[a, b]$ contain all the points $\lambda_j(t)$, $j = 1, 2, \ldots, d$. Let $\varphi_l \rightrightarrows f'$ on $[a, b]$ as $l \to \infty$, where φ_l is a polynomial, $l \in \mathbb{N}$. Then, $f_l(\lambda) := f(a) + \int_a^\lambda \varphi_l(t)\,dt \rightrightarrows f(\lambda)$ on $[a, b]$. Applying the estimate (3.1.12) gives

$$\left\| \frac{f(X(t+s)) - f(X(t))}{s} - \frac{f_l(X(t+s)) - f_l(X(t))}{s} \right\|$$

$$\leqslant \sqrt{d}\, \|f' - f_l'\|_\infty \left\| \frac{X(t+s) - X(t)}{s} \right\|.$$

By the estimate (3.1.7),

$$\left\| T_{f^{[1]}}^{X(t),X(t)}(X(t)) - T_{f_l^{[1]}}^{X(t),X(t)}(X(t)) \right\| \leqslant \sqrt{d}\, \|f' - f_l'\|_\infty \|X(t)\|.$$

Combining the latter two inequalities with (5.3.1) for polynomials completes the proof. □

The formula (5.3.1) is an excellent instrument for studying the derivatives of operator functions. In particular, by (3.1.6) and the formula (5.3.1), we obtain the following estimate for the derivative:

$$\left\| \frac{df(X(t))}{dt} \right\|_2 \leqslant \max_{a \leqslant \lambda \leqslant b} |f'(\lambda)| \left\| \frac{dX(t)}{dt} \right\|_2. \tag{5.3.2}$$

In this section we present more delicate estimates for operator derivatives which extend and complement (5.3.2) to the case of other norms.

Below we prove extension of (5.3.1) to the higher order case.

Theorem 5.3.2 *Let* $H, V \in \mathcal{B}_{sa}(\ell_d^2)$. *If* $f \in C^k(\mathbb{R})$, $k \in \mathbb{N}$, *then the function* $t \mapsto f(H + tV)$ *is differentiable k times in the operator norm and*

$$\frac{d^k f(H + tV)}{dt^k} = k! \, T_{f^{[k]}}^{H+tV,\dots,H+tV} (\underbrace{V, \dots, V}_{k}). \tag{5.3.3}$$

Proof We proceed by induction and, for simplicity of exposition, calculate only the derivative at $t = 0$. The base of induction is proved in Proposition 5.3.1, so we only need to confirm the inductive step.

Assume that the function $t \to f(H+tV)$ is differentiable $k - 1$ times and (5.3.3) holds for $k - 1$. Thus, we have

$$\frac{d^k}{dt^k} f(H + tV)\Big|_{t=0}$$

$$= \lim_{t \to 0} \frac{(k-1)!}{t} \left(T_{f^{[k-1]}}^{H+tV,\dots,H+tV}(V, \dots, V) - T_{f^{[k-1]}}^{H,\dots,H}(V, \dots, V) \right)$$

$$= \lim_{t \to 0} \frac{(k-1)!}{t} \sum_{j=0}^{k-1} \left(T_{f^{[k-1]}}^{\overbrace{H+tV,\dots,H+tV}^{k-j},\overbrace{H,\dots,H}^{j}}(V, \dots, V) \right.$$

$$\left. - T_{f^{[k-1]}}^{\overbrace{H+tV,\dots,H+tV}^{k-j-1},\overbrace{H,\dots,H}^{j+1}}(V, \dots, V) \right). \tag{5.3.4}$$

By the perturbation formula (4.1.9),

$$\frac{1}{t}\left(T_{f^{[k-1]}}^{\overbrace{H+tV,\ldots,H+tV}^{k-j},\overbrace{H,\ldots,H}^{j}} - T_{f^{[k-1]}}^{\overbrace{H+tV,\ldots,H+tV}^{k-j-1},\overbrace{H,\ldots,H}^{j+1}}\right)$$

$$\times (V,\ldots,V)$$

$$= T_{f^{[k]}}^{\overbrace{H+tV,\ldots,H+tV}^{k-j},\overbrace{H,\ldots,H}^{j+1}}\,(V,\ldots,V). \tag{5.3.5}$$

By Proposition 4.1.6,

$$\lim_{t\to 0} T_{f^{[k]}}^{H+tV,\ldots,H+tV,H,\ldots,H}(V,\ldots,V) = T_{f^{[k]}}^{H,\ldots,H}(V,\ldots,V). \tag{5.3.6}$$

Combining (5.3.4)–(5.3.6) completes the proof of (5.3.3).   □

The function in Theorem 5.3.2 is, in fact, k times continuously Fréchet differentiable at every $H \in \mathcal{B}_{sa}(\ell_d^2)$. This fact is proved in [88, Theorem 2.3.1] by direct computation for polynomials and induction along with performing an approximation of general functions by polynomials that is based on coarse Hilbert-Schmidt bounds for multilinear Schur multipliers, where the constant grows polynomially with the dimension of the matrix (see Sect. 4.1.3 for details about the bound). Ideologically this is similar to but technically different from the proof of Theorem 5.3.2. In Theorem 5.3.2, we use a general approximation method of Proposition 4.1.6 and perform the induction step based on purely algebraic Proposition 4.1.5. More general results on Fréchet differentiability of operator functions in the infinite dimensional setting along with delicate bounds for Fréchet derivatives are discussed in Sect. 5.3.4.

5.3.2 Differentiation in $\mathcal{B}(\mathcal{H})$ Along Multiplicative Paths of Unitaries

To prove an analog of (5.3.3) in the infinite-dimensional unitary case, we need the following lemma obtained in [168, Lemma 2.6 and (3.2)].

Lemma 5.3.3 Let $f \in B_{\infty 1}^n(\mathbb{T})$, let A and $U \in \mathcal{B}(\mathcal{H})$ be self-adjoint and unitary operator, respectively. Denote $U(t) = e^{itA}U$, $t \in \mathbb{R}$. Then, for all $1 \leqslant k \leqslant n-1$ and all $j_1,\ldots,j_k \in \mathbb{N}$,

$$\frac{d}{dt} T_{f^{[k]}}^{\tilde{U}_{k+1}(t)}(A^{j_1}U(t),\ldots,A^{j_k}U(t))\Big|_{t=s}$$

$$= i\sum_{m=1}^{k+1} T_{f^{[k+1]}}^{\tilde{U}_{k+2}(s)}(A^{j_1}U(s),\ldots,A^{j_{m-1}}U(s),AU(s),A^{j_m}U(s),\ldots,A^{j_k}U(s))$$

$$+ i \sum_{m=1}^{k} T_{f^{[k]}}^{\tilde{U}_{k+1}(s)}$$

$$\times (A^{j_1}U(s), \ldots, A^{j_{m-1}}U(s), A^{j_m+1}U(s), A^{j_{m+1}}U(s), \ldots, A^{j_k}U(s)),$$

where $\tilde{U}_r(t)$ is the tuple consisting of r copies of $U(t)$.

The following theorem is proved for $n = 1$ in [141, (5)] and for a general n in [168, Theorem 3.1], with precise coefficients found in [205].

Theorem 5.3.4 *Let* $f \in B_{\infty 1}^n(\mathbb{T})$, *let* $A \in \mathcal{B}_{sa}(\mathcal{H})$ *and let* $U \in \mathcal{B}(\mathcal{H})$ *be a unitary operator. Set* $U(t) = e^{itA}U, t \in \mathbb{R}$. *Then,*

$$\frac{1}{k!} \frac{d^k}{dt^k} f(U(t))\Big|_{t=s} = i^k \sum_{l=1}^{k} \sum_{\substack{j_1,\ldots,j_l \geqslant 1 \\ j_1+\cdots+j_l=k}} \frac{1}{j_1! \ldots j_l!} T_{f^{[l]}}^{\tilde{U}_{l+1}(s)}(A^{j_1}U(s), \ldots, A^{j_l}U(s)),$$

$$(5.3.7)$$

for all $1 \leqslant k \leqslant n-1$. *Moreover,*

$$\left\| \frac{d^n}{dt^n} f(U(t))\Big|_{t=s} \right\| \leqslant c_n \|f\|_{B_{\infty 1}^n(\mathbb{T})} \|A\|^n. \qquad (5.3.8)$$

Proof The formula (5.3.7) in the case $k = 1$ was established in [141, (5)].

The formula (5.3.7) for an arbitrary $1 \leqslant k \leqslant n-1$ can be established by induction. We assume that it holds for $k = p < n - 1$ and verify below that it holds for $k = p + 1$. Denote $V_{j,t} := A^j U(t)$. Applying Lemma 5.3.3 gives

$$\frac{d^{p+1}}{dt^{p+1}} f(U(t))\Big|_{t=s} = i^p \sum_{l=1}^{p} \sum_{\substack{j_1,\ldots,j_l \geqslant 1 \\ j_1+\cdots+j_l=p}} \frac{p!}{j_1! \ldots j_l!} \frac{d}{dt} T_{f^{[l]}}^{\tilde{U}_{l+1}(t)}(V_{j_1,t}, \ldots, V_{j_l,t})\Big|_{t=s}$$

$$= i^{p+1} \sum_{l=1}^{p} \sum_{\substack{j_1,\ldots,j_l \geqslant 1 \\ j_1+\cdots+j_l=p}} \frac{p!}{j_1! \ldots j_l!}$$

$$\times \sum_{m=1}^{l} T_{f^{[l]}}^{\tilde{U}_{l+1}(t)}(V_{j_1,t}, \ldots, V_{j_{m-1},t}, V_{j_m+1,t}, V_{j_{m+1},t}, \ldots, V_{j_l,t})$$

$$+ i^{p+1} \sum_{l=1}^{p} \sum_{\substack{j_1,\ldots,j_l \geqslant 1 \\ j_1+\cdots+j_l=p}} \frac{p!}{j_1! \ldots j_l!}$$

$$\times \sum_{m=1}^{l+1} T_{f^{[l+1]}}^{\tilde{U}_{l+2}(t)}(V_{j_1,t}, \ldots, V_{j_{m-1},t}, V_{1,t}, V_{j_m,t}, \ldots, V_{j_l,t})$$

$$=: i^{p+1}(S_1 + S_2),$$

where $\tilde{U}_{k+1}(t)$ is the tuple consisting of $k+1$ copies of $U(t)$. Making the substitution $i_r = j_r$, $1 \leqslant r \neq m \leqslant l$ and $i_m = j_m + 1$ in S_1, we obtain

$$S_1 = \sum_{l=1}^{p} \sum_{m=1}^{l} \sum_{\substack{i_r \geqslant 1, r \neq m, i_m \geqslant 2 \\ i_1 + \cdots + i_l = p+1}} \frac{p!}{i_1! \ldots i_{m-1}!(i_m - 1)!i_{m+1}! \ldots i_l!}$$

$$\times T_{f^{[l]}}^{\tilde{U}_{l+1}(t)}(V_{i_1,t}, \ldots, V_{i_l,t}).$$

Relabeling the summands of S_2 via the mapping $l \mapsto l - 1$ and making the substitution $i_r = j_r$, $1 \leqslant r \leqslant m - 1$, $i_m = 1$ and $i_r = j_{r-1}$, $m + 1 \leqslant r \leqslant l$, we obtain

$$S_2 = \sum_{l=2}^{p+1} \sum_{\substack{j_1, \ldots, j_{l-1} \geqslant 1 \\ j_1 + \cdots + j_{l-1} = p}} \frac{p!}{j_1! \ldots j_{l-1}!}$$

$$\times \sum_{m=1}^{l} T_{f^{[l]}}^{\tilde{U}_{l+1}(t)}(V_{j_1,t}, \ldots, V_{j_{m-1},t}, V_{1,t}, V_{j_m,t}, \ldots, V_{j_{l-1},t})$$

$$= \sum_{l=2}^{p+1} \sum_{m=1}^{l} \sum_{\substack{i_1, \ldots, i_l \geqslant 1, i_m = 1 \\ i_1 + \cdots + i_l = p+1}} \frac{p!}{i_1! \ldots i_{m-1}! i_{m+1}! \ldots i_l!} T_{f^{[l]}}^{\tilde{U}_{l+1}(t)}(V_{i_1,t}, \ldots, V_{i_l,t}).$$

Thus,

$$S_1 + S_2 = T_{f^{[1]}}^{\tilde{U}_2(t)}(V_{p+1,t})$$

$$+ \sum_{l=2}^{p} \sum_{m=1}^{l} \Bigg(\sum_{\substack{i_r \geqslant 1, r \neq m, i_m \geqslant 2 \\ i_1 + \cdots + i_l = p+1}} + \sum_{\substack{i_1, \ldots, i_l \geqslant 1, i_m = 1 \\ i_1 + \cdots + i_l = p+1}} \Bigg) \frac{p!}{i_1! \ldots i_{m-1}!(i_m - 1)!i_{m+1}! \ldots i_l!}$$

$$\times T_{f^{[l]}}^{\tilde{U}_{l+1}(t)}(V_{i_1,t}, \ldots, V_{i_l,t})$$

$$+ \sum_{m=1}^{p+1} \sum_{\substack{i_1, \ldots, i_{p+1} \geqslant 1, i_m = 1 \\ i_1 + \cdots + i_{p+1} = p+1}} \frac{p!}{i_1! \ldots i_{m-1}!(i_m - 1)!i_{m+1}! \ldots i_{p+1}!}$$

$$\times T_{f^{[p+1]}}^{\tilde{U}_{p+2}(t)}(V_{i_1,t}, \ldots, V_{i_{p+1},t}),$$

so

$$S_1 + S_2 = T^{\tilde{U}_2(t)}_{f^{[1]}}(V_{p+1,t})$$

$$+ \sum_{l=2}^{p} \sum_{m=1}^{l} \sum_{\substack{i_1,\ldots,i_l \geqslant 1 \\ i_1+\cdots+i_l=p+1}} \frac{p!}{i_1!\ldots i_{m-1}!(i_m-1)!i_{m+1}!\ldots i_l!} T^{\tilde{U}_{l+1}(t)}_{f^{[l]}}(V_{i_1,t},\ldots,V_{i_l,t})$$

$$+ (p+1)!\, T^{\tilde{U}_{p+2}(t)}_{f^{[p+1]}}(V_{1,t},\ldots,V_{1,t}).$$

Since

$$\sum_{m=1}^{l} \frac{p!}{i_1!\ldots i_{m-1}!(i_m-1)!i_{m+1}!\ldots i_l!} = p!\,\frac{i_1+\cdots+i_l}{i_1!\ldots i_l!} = \frac{(p+1)!}{i_1!\ldots i_l!},$$

it follows that

$$S_1 + S_2 = T^{\tilde{U}_2(t)}_{f^{[1]}}(V_{p+1,t}) + \sum_{l=2}^{p} \sum_{\substack{i_1,\ldots,i_l \geqslant 1 \\ i_1+\cdots+i_l=p+1}} \frac{(p+1)!}{i_1!\ldots i_l!} T^{\tilde{U}_{l+1}(t)}_{f^{[l]}}(V_{i_1,t},\ldots,V_{i_l,t})$$

$$+ (p+1)!\, T^{\tilde{U}_{p+2}(t)}_{f^{[p+1]}}(V_{1,t},\ldots,V_{1,t})$$

$$= \sum_{l=1}^{p+1} \sum_{\substack{i_1,\ldots,i_l \geqslant 1 \\ i_1+\cdots+i_l=p+1}} \frac{(p+1)!}{i_1!\ldots i_l!} T^{\tilde{U}_{l+1}(t)}_{f^{[l]}}(V_{i_1,t},\ldots,V_{i_l,t}),$$

proving (5.3.7).

The estimate (5.3.8) follows from the estimates (4.3.34) for the multiple operator integral and (4.3.35) for its symbol. □

5.3.3 Differentiation in $\mathcal{B}(\mathcal{H})$ and $\mathcal{S}^1$ Along Linear Paths of Self-adjoints

Differentiability in the infinite-dimensional setting imposes stronger restrictions on the respective class of scalar functions than in the finite-dimensional setting. In particular, the condition $f \in C^1(\mathbb{R})$ is not sufficient for differentiability of operator functions in the infinite-dimensional case even when the involved operators are bounded [141, Theorem 8] (see also [80, 81]). It is proved in [143, Theorem 2] that if f is an element of the Besov space $B^1_{\infty 1}(\mathbb{R})$, then f is Gâteaux differentiable with respect to the $\mathcal{S}^p$-norm, $1 \leqslant p < \infty$, and the operator norm at every self-adjoint operator. This restriction on f is relaxed in [19] by methods of complex analysis (see [9, Subsection 3.13] for details). It also follows (see theorem below) that $f \in B^1_{\infty 1}(\mathbb{R})$ is Fréchet S^p-differentiable, $1 \leqslant p < \infty$, at every self-adjoint operator and Fréchet differentiable with respect to the operator norm at every bounded self-adjoint operator, and in the case $p \neq 1$ the restriction on f is substantially relaxed in the next section. A necessary condition for differentiability of f in the S^1-norm, which creates a small gap with the sufficient condition, is obtained in [143, Theorem 3] via nuclearity criterion for Hankel operators. Higher order Gâteaux differentiability of operator functions with respect to the operator and trace class norms is established in [146] and higher order Fréchet differentiability in [119, Theorem 4.1]. The aforementioned sufficient conditions for S^1- and $\mathcal{B}(\mathcal{H})$-differentiability are summarised and proved in the theorem below.

Theorem 5.3.5 Let $A \in \mathcal{D}_{sa}$, $X = X^* \in S^1$ (or $\mathcal{B}(\mathcal{H})$), $n \in \mathbb{N}$, and $f \in B^1_{\infty 1}(\mathbb{R}) \cap B^n_{\infty 1}(\mathbb{R})$. Then, the function f is differentiable n times at A along the direction X in the sense of Gâteaux in the S^1-norm (or operator norm),

$$\frac{d^n}{dt^n} f(A + tX)\Big|_{t=0} = n! \, T^{A,\ldots,A}_{f^{[n]}}(X, \ldots, X), \qquad (5.3.9)$$

and

$$\left\| \frac{d^n}{dt^n} f(A + tX)\Big|_{t=0} \right\|_p \leqslant c_n \|f\|_{B^n_{\infty 1}(\mathbb{R})} \|X\|_p^n \qquad (5.3.10)$$

for $1 \leqslant p \leqslant \infty$. Moreover, the function f is n times continuously Fréchet differentiable at A in the S^1-norm (or, if A is bounded, in the operator norm).

Proof We note that by (4.3.12) and Theorem 4.3.4, $T^{A,\ldots,A}_{f^{[k]}}(X, \ldots, X)$ is well defined for every $k = 1, \ldots, n$ because $B^1_{\infty 1}(\mathbb{R}) \cap B^n_{\infty 1}(\mathbb{R}) \subset \cap_{k=1}^n B^k_{\infty 1}(\mathbb{R})$. The proof goes by induction on n.

By Theorems 3.3.8 and 3.3.14,

$$\frac{f(A + tX) - f(A)}{t} = T^{A+tX,A}_{f^{[1]}}(X).$$

Applying Proposition 3.3.9(i), the estimate (3.3.6), and Theorem 3.3.8 completes the proof in the case $n = 1$.

The proof of existence of the higher order derivatives follows the same steps as the proof of (5.3.2), with replacement of the perturbation formula and continuity of a multiple operator integral by their infinite dimensional counterparts stated in Propositions 4.3.14 and 4.3.15(i). We note that Proposition 4.3.15 is applicable thanks to Theorem 4.3.4.

Now we justify the continuous Fréchet differentiability. The first order Fréchet differentiability follows from the representation (5.3.9) and Propositions 2.11.3

and 3.3.9(i). Applying the connection between the Fréchet differential and Gâteaux derivative (2.11.2) along with (5.3.9) gives

$$D_1^1 f(A + X)(X_1) - D_1^1 f(A)(X)$$
$$= \left(T_{f^{[1]}}^{A+X,A+X}(X_1) - T_{f^{[1]}}^{A+X,A}(X_1)\right) + \left(T_{f^{[1]}}^{A+X,A}(X_1) - T_{f^{[1]}}^{A,A}(X_1)\right).$$

Applying Proposition 3.3.9(i) confirms continuity of the Fréchet differential $D_1^1 f(A)$. The proof of a higher order continuous Fréchet differentiability is technically more involved, and we refer the reader to [119, Theorem 4.1] for details. □

Every operator differentiable function arises from a sufficiently smooth scalar function.

Theorem 5.3.6 *Let $f \in C(\mathbb{R})$. If $t \to f(A + tX) - f(A)$ is differentiable as a function on $\mathbb{R}$ to the space $\mathcal{B}(\mathcal{H})$ equipped with the strong operator topology for all $A \in \mathcal{D}_{sa}$ and $X \in \mathcal{B}_{sa}(\mathcal{H})$ for every separable Hilbert space $\mathcal{H}$, then f is operator Lipschitz on $\mathbb{R}$ and, in particular, $f \in B_{11}^1(\mathbb{R})_{loc}$.*

Proof The result follows from combination of [9, Theorem 1.2.4] and Theorem 5.1.6(ii). □

Remark 5.3.7 Although existence of derivatives of operator functions for *all* operators A and X requires smoothness of the respective scalar function, operator derivatives at a particular operator point A along a particular direction X can exist for nonsmooth scalar functions [18]. We also refer the reader to Sect. 5.3.7, where the connection between operator Lipschitzness and operator differentiability is discussed.

5.3.4 Differentiation in $\mathcal{S}^p$ Along Linear Paths of Self-adjoints

In this section we will see that the set of differentiable functions with respect to the $\mathcal{S}^p$-norm, $1 < p < \infty$, is larger than the set of $\mathcal{S}^1$-differentiable functions.

The following result is due to [110, Theorems 7.15, 7.17, 7.18].

Theorem 5.3.8 *Let $1 < p < \infty$.*

(i) *A function $f : \mathbb{R} \to \mathbb{R}$ is Gâteaux $\mathcal{S}^p$-differentiable at all $A \in \mathcal{B}_{sa}(\mathcal{H})$ if and only if f is differentiable on $\mathbb{R}$ and has bounded derivative on all compact subsets of $\mathbb{R}$. Moreover, every differentiable function $f : \mathbb{R} \to \mathbb{R}$ with $f' \in C_b(\mathbb{R})$ is Gâteaux $\mathcal{S}^p$-differentiable at every $A \in \mathcal{D}_{sa}$ along $\mathcal{S}^p_{sa}$. The respective Gâteaux derivative is given by $D_{G,p} f(A) = T_{f^{[1]}}^{A,A}$.*

(ii) *A function $f : \mathbb{R} \to \mathbb{R}$ is Fréchet $\mathcal{S}^p$-differentiable at all $A \in \mathcal{B}_{sa}(\mathcal{H})$ if and only if $f \in C^1(\mathbb{R})$. The respective Fréchet differential is given by $D_p f(A) = T_{f^{[1]}}^{A,A}$.*

It is proved in [24, Theorem 5.5] that a function f is n times Fréchet $\mathcal{S}^p$-differentiable at every self-adjoint operator if f is in the Wiener class $W_{n+1}(\mathbb{R})$, which is improved in Theorem 5.3.5. It is proved in [55, Theorem 4.1] that a function f in $C^n(\mathbb{R})$ is n times Gâteaux $\mathcal{S}^2$-differentiable at every bounded self-adjoint operator and, under the additional assumption "$f^{(j)}$ is bounded, $j = 0, \ldots, n$", at every self-adjoint operator. We state below strengthening of these results due to [119, Theorems 3.4, 3.6, 3.7(ii)].

Theorem 5.3.9 *Let* $1 < p < \infty$, $n \in \mathbb{N}$. *The following assertions hold.*

(i) *Let* $f \in C^n(\mathbb{R})$ *satisfy* $f', \ldots, f^{(n-1)} \in C_b(\mathbb{R})$ *and* $f^{(n)} \in C_0(\mathbb{R})$. *Then* f *is* n *times continuously Fréchet* $\mathcal{S}^p$-*differentiable at every* $A \in \mathcal{D}_{sa}$ *and*

$$D_p^k f(A)(X_1, \ldots, X_k) = \sum_{\sigma \in \text{Sym}_k} T_{f^{[k]}}^{A, \ldots, A}(X_{\sigma(1)}, \ldots, X_{\sigma(k)}), \qquad (5.3.11)$$

for every $k = 1, \ldots, n$, *and all* $X_1, \ldots, X_k \in \mathcal{S}^p$, *where* $T_{f^{[k]}}^{A, \ldots, A}$ *is given by Definition 4.3.3 and* Sym_k *denotes the group of all permutations of the set* $\{1, \ldots, k\}$.

(ii) *Let* $f : \mathbb{R} \to \mathbb{C}$ *be a locally Lipschitz function. Then* f *is* n *times continuously Fréchet* $\mathcal{S}^p$-*differentiable at every* $A \in \mathcal{B}_{sa}(\mathcal{H})$ *and* (5.3.11) *holds if and only if* $f \in C^n(\mathbb{R})$.

(iii) *Let* $f \in C^n(\mathbb{R})$ *satisfy* $f', \ldots, f^{(n)} \in C_b(\mathbb{R})$. *Then,* f *is* $n - 1$ *times continuously Fréchet* $\mathcal{S}^p$-*differentiable and* n *times Gâteaux* $\mathcal{S}^p$-*differentiable at every* $A \in \mathcal{D}_{sa}$, *with*

$$D_{G,p}^n f(A)(X) = n!\, T_{f^{[n]}}^{A, \ldots, A}(X, \ldots, X)$$

for all $X = X^* \in \mathcal{S}^p$.

We note that the results of Theorems 4.3.10 and 4.3.13 play a crucial role in the proof of Theorem 5.3.9, but omit the proof. We also note that the assumption "$f \in C^1(\mathbb{R})$ and f' is bounded" is not sufficient for Fréchet differentiability of f at an arbitrary unbounded operator, as demonstrated in [110, Example 7.20].

We also have the following characterization of n times Gâteaux $\mathcal{S}^p$-differentiable functions extending the respective result of Theorem 5.3.8. The sufficient conditions are established [52] and the necessary conditions in [119, Proposition 3.9].

Theorem 5.3.10 *Let* $1 < p < \infty$ *and* $n \in \mathbb{N}$.

(i) *Let* $f : \mathbb{R} \to \mathbb{C}$ *be a locally Lipschitz function. Then,* f *is* n *times Gâteaux* $\mathcal{S}^p$-*differentiable at every* $A \in \mathcal{B}_{sa}(\mathcal{H})$ *if and only if* f *is* n *times differentiable on* $\mathbb{R}$ *and* $f^{(n)}$ *is bounded on all compact subsets of* $\mathbb{R}$.

(ii) *Let* $f : \mathbb{R} \to \mathbb{C}$ *be a Lipschitz function. Then,* f *is* n *times Gâteaux* $\mathcal{S}^p$-*differentiable at every* $A \in \mathcal{D}_{sa}$ *if and only if* f *is* n *times differentiable on* $\mathbb{R}$ *and* $f', \ldots, f^{(n)}$ *are bounded.*

Problem 5.3.11 Find criteria for higher order Fréchet S^p-differentiability, $1 \leqslant p \leqslant \infty$, of an operator function $f(A)$, $A \in \mathcal{D}_{sa}$, in terms of smoothness properties of the respective scalar function $f : \mathbb{R} \to \mathbb{C}$.

The results below improve the bound (5.3.10) in the case $1 < p < \infty$ and, when the norm $\|\cdot\|_1 = \mathrm{Tr}(|\cdot|)$ is replaced with the smaller seminorm $|\mathrm{Tr}(\cdot)|$, in the case $p = 1$.

Theorem 5.3.12 *Let* $A \in \mathcal{D}_{sa}$, $X \in \mathcal{B}_{sa}(\mathcal{H})$, $n \in \mathbb{N}$ *and* $f \in C^n(\mathbb{R})$. *If* A *is unbounded assume also that* $f', \ldots, f^{(n-1)} \in C_b(\mathbb{R})$ *and* $f^{(n)} \in C_0(\mathbb{R})$. *The following assertions hold.*

(i) If $X \in S^p$, $1 < p < \infty$, *then*

$$\left\| \frac{d^n}{dt^n} f(A + tX) \big|_{t=0} \right\|_p \leqslant c_{p,n} \| f^{(n)} \|_\infty \| X \|_{pn}^n \leqslant c_{p,n} \| f^{(n)} \|_\infty \| X \|_p^n.$$

(5.3.12)

(ii) If $X \in S^n$, *then*

$$\left| \mathrm{Tr} \left(\frac{d^n}{dt^n} f(A + tX) \big|_{t=0} \right) \right| \leqslant c_n \| f^{(n)} \|_\infty \| X \|_n^n.$$

(5.3.13)

Moreover, if A *is bounded, then* $\| f^{(n)} \|_\infty$ *in* (5.3.12) *and* (5.3.13) *can be replaced with* $\max_{t \in \sigma(A)} | f^{(n)}(t) |$.

Proof

(i) It follows from Theorem 4.3.11 and (5.3.11) that

$$\left\| \frac{d^n}{dt^n} f(A + tX) \big|_{t=0} \right\|_p \leqslant c_{p,n} \| f^{(n)} \|_\infty \| X \|_{pn}^n.$$

Since $\| \cdot \|_{pn} \leqslant \| \cdot \|_p$, the inequality (5.3.12) follows.
(ii) We note that

$$\mathrm{Tr} \left(\frac{d^n}{dt^n} f(A + tX) \big|_{t=0} \right) = \mathrm{Tr} \left(T_\psi^{A, \ldots, A} (\underbrace{X, \ldots, X}_{n-1}) X \right),$$

(5.3.14)

where

$$\psi(\lambda_0, \lambda_1, \ldots, \lambda_{n-1}) = f^{[n]}(\lambda_0, \lambda_0, \lambda_1, \ldots, \lambda_{n-1}).$$

The result follows upon applying Hölder's inequality and Theorem 4.3.10 in (5.3.14). More details can be found in [163, p. 536]. $\square$

5.3.5 Differentiation of Functions of Contractive and Dissipative Operators

Let $n \in \mathbb{N}$ and let $\mathbb{P}_+$ be the Riesz projection on $B^n_{\infty 1}(\mathbb{T})$, that is,

$$\mathbb{P}_+(f) : z \to \sum_{n \geqslant 0} \mathcal{F} f(n) z^n, \quad f \in B^n_{\infty 1}(\mathbb{T}).$$

The operator $\mathbb{P}_+$ is bounded on $B^n_{\infty 1}(\mathbb{T})$. Define

$$\left(B^n_{\infty 1}(\mathbb{T}) \right)_+ := \mathbb{P}_+ B^n_{\infty 1}(\mathbb{T}).$$

The functions in $\left(B^n_{\infty 1}(\mathbb{T}) \right)_+$ admit a natural extension on the closed unit disc $\bar{\mathbb{D}}$ and they are analytic in $\mathbb{D}$.

The following result is established in [147, Theorem 5.3].

Theorem 5.3.13 *Let A be a contraction on $\mathcal{H}$ and let $X \in \mathcal{B}(\mathcal{H})$ be such that $A + X$ is a contraction. Let $n \in \mathbb{N}$ and $f \in \left(B^n_{\infty 1}(\mathbb{T}) \right)_+$. Then, the function f is differentiable n times at A along the direction X in the sense of Gâteaux in the operator norm,*

$$\frac{d^n}{dt^n} f(A + tX) \Big|_{t=0} = n! \, T^{A,\ldots,A}_{f^{[n]}}(X, \ldots, X),$$

and

$$\left\| \frac{d^n}{dt^n} f(A + tX) \Big|_{t=0} \right\| \leqslant c_n \| f \|_{B^n_{\infty 1}(\mathbb{T})} \| X \|^n.$$

The next estimate follows from Theorems 4.3.19 and 5.3.13.

Theorem 5.3.14 *Let A be a contraction on $\mathcal{H}$ and let $X \in \mathcal{S}^p$, $1 < p < \infty$, be such that $A + X$ is a contraction. Let $n \in \mathbb{N}$ and let $f \in C^n(\mathbb{T})$ satisfy $\sum_{j=0}^{\infty} |j|^n |\mathcal{F} f(j)| < \infty$. Then,*

$$\left\| \frac{d^n}{dt^n} f(A + tX) \Big|_{t=0} \right\|_p \leqslant c_{p,n} \| f^{(n)} \|_\infty \| X \|_p^n.$$

The next result is due to [147, Theorem 6.1].

Theorem 5.3.15 *Let A be a contraction on $\mathcal{H}$ and let $X \in \mathcal{B}(\mathcal{H})$ be such that $A + X$ is a contraction. Let f be a function analytic in $\mathbb{D}$ such that $f' \in \mathcal{A}(\mathbb{D})$. Then, the function f is Gâteaux differentiable at A along the direction X in the Hilbert-Schmidt norm and $D_{G,2} f(A) = T^{A,A}_{f^{[1]}}$.*

Let $\left(B_{\infty 1}^n(\mathbb{R})\right)_+$ denote the set of those $f \in B_{\infty 1}^n(\mathbb{R})$ for which $\operatorname{supp} \mathcal{F} f \subset [0, \infty)$. The next result is obtained in [6, Theorem 9.1].

Theorem 5.3.16 *Let A be a maximal dissipative operator densely defined in $\mathcal{H}$ and let $X \in \mathcal{B}(\mathcal{H})$ be such that $A + X$ is a maximal dissipative operator. Let $n \in \mathbb{N}$ and $f \in \left(B_{\infty 1}^n(\mathbb{R})\right)_+$. Then, the function f is n times Gâteaux differentiable at A along the direction X in the operator norm,*

$$\frac{d^n}{dt^n} f(A + tX)\Big|_{t=0} = n!\, T_{f^{[n]}}^{A, \ldots, A}(X, \ldots, X),$$

and

$$\left\| \frac{d^n}{dt^n} f(A + tX)\Big|_{t=0} \right\| \leqslant c_n \|f\|_{B_{\infty 1}^n(\mathbb{R})} \|X\|^n.$$

5.3.6 Differentiation in Noncommutative L^p-Spaces

Differentiability of operator functions in noncommutative L^p-spaces and general noncommutative symmetric spaces is studied in [24, 63].

The following results are obtained in [63, Theorem 5.16, Corollary 7.3] for noncommutative symmetric spaces $\mathcal{E}$ of a semifinite von Neumann algebra, including the noncommutative L^p-spaces, $1 \leqslant p \leqslant \infty$.

Theorem 5.3.17 *Let $(\mathcal{M}, \tau)$ be a semifinite von Neumann algebra, let E be a separable symmetric Banach function space on $(0, \infty)$ and $\mathcal{E} = E(\mathcal{M}, \tau)$ the respective noncommutative symmetric space. If f is such that $f^{[1]} \in \mathfrak{A}_1$, then f is Gâteaux operator differentiable along $\mathcal{E}_{sa}$ at every point $A \eta \mathcal{M}_{sa}$ and the Gâteaux derivative with respect to the $\mathcal{E}$-norm equals $D_{G, \mathcal{E}} f(A) = T_{f^{[1]}}^{A, A}$.*

Corollary 5.3.18 *If $f \in B_{\infty 1}^1(\mathbb{R})$, then f is Gâteaux operator differentiable along $\mathcal{E}_{sa}$ at every point $A \eta \mathcal{M}_{sa}$ for every semifinite von Neumann algebra $(\mathcal{M}, \tau)$ and noncommutative symmetric space $\mathcal{E} = E(\mathcal{M}, \tau)$.*

A sufficient condition for differentiability in noncommutative L^p-spaces, $1 < p < \infty$, that does not involve a separation of variables of $f^{[1]}$ is established in [63, Corollary 6.10].

Theorem 5.3.19 *If $f : \mathbb{R} \to \mathbb{R}$ is continuously differentiable and f' is of bounded variation, then f is Gâteaux operator differentiable along $L^p(\mathcal{M}, \tau)_{sa}$, $1 < p < \infty$, at every point $A \eta \mathcal{M}_{sa}$ and the Gâteaux derivative with respect to the $L^p(\mathcal{M}, \tau)$-norm equals $D_{G, p} f(A) = T_{f^{[1]}}^{A, A}$.*

A necessary and sufficient condition for Gâteaux differentiability in the context of an arbitrary semifinite von Neumann algebra is known only for noncommutative L^2-spaces, and it is due to [63, Proposition 6.11].

Theorem 5.3.20 *A Lipschitz function* $f : \mathbb{R} \to \mathbb{R}$ *is Gâteaux differentiable along every self-adjoint direction in* $L^2(\mathcal{M}, \tau)$ *at every point* $A \eta \mathcal{M}_{sa}$ *for every semifinite von Neumann algebra* $(\mathcal{M}, \tau)$ *if and only if* f *is continuously differentiable. In this case the Gâteaux derivative with respect to the* $L^2(\mathcal{M}, \tau)$-*norm equals* $D_{G,2}f(A) = T_{f^{[1]}}^{A,A}$.

Higher order Fréchet differentiability in the ideal $\mathcal{L}^p(\mathcal{M}, \tau) = L^p(\mathcal{M}, \tau) \cap \mathcal{M}$, $1 \leqslant p < \infty$, of a semifinite von Neumann algebra $(\mathcal{M}, \tau)$ is obtained in [24, Theorem 5.7] for a more restrictive class of functions than the one found for the first order differentiability. In fact, the latter result holds for more general symmetrically normed ideals. We state this result below.

Theorem 5.3.21 *Let* $\mathcal{M}$ *be a semifinite von Neumann algebra acting on a separable Hilbert space* $\mathcal{H}$ *and* $\mathcal{I}$ *a symmetrically normed ideal of* $\mathcal{M}$ *with property* (F). *Let* $n \in \mathbb{N}$. *Then, every* $f \in W_{n+1}(\mathbb{R})$ *is* n *times continuously Fréchet differentiable at every* $A \eta \mathcal{M}_{sa}$ *and*

$$D_{\mathcal{I}}^n f(A)(X_1, \ldots, X_n) = \sum_{\sigma \in \mathrm{Sym}_n} T_{f^{[n]}}^{A, \ldots, A}(X_{\sigma(1)}, \ldots, X_{\sigma(n)}).$$

The following problem generalizes Problem 5.3.11.

Problem 5.3.22 Find criteria for an arbitrary order Gâteaux and Fréchet differentiability of operator functions in noncommutative symmetric spaces.

5.3.7   Gâteaux and Fréchet $\mathcal{I}$-Differentiable Functions

Let $\mathcal{I}$ be a symmetrically normed (s. n.) ideal of $\mathcal{B}(\mathcal{H})$, including the case $\mathcal{I} = \mathcal{B}(\mathcal{H})$. Denote by $\mathcal{I}_{sa}$ the set of self-adjoint elements in $\mathcal{I}$.

A function $f \in C(\mathbb{R})$ is called *Gâteaux* $\mathcal{I}$-*differentiable at* $A \in \mathcal{B}_{sa}(\mathcal{H})$ *along* $\mathcal{I}_{sa}$ if there is a bounded linear operator $D_G^{\mathcal{I}} f(A)$ from $\mathcal{I}_{sa}$ to $\mathcal{I}$ such that

$$\lim_{t \to 0} \frac{1}{t} \left\| f(A + tX) - f(A) - t D_G^{\mathcal{I}} f(A)(X) \right\|_{\mathcal{I}} = 0$$

for each $X \in \mathcal{I}_{sa}$ and $t \in \mathbb{R}$. The operator $D_G^{\mathcal{I}} f(A)$ is called the Gâteaux $\mathcal{I}$-derivative of f at A. If f is real-valued then $D_G^{\mathcal{I}} f(A) \colon \mathcal{I}_{sa} \to \mathcal{I}_{sa}$.

A function $f \in C(\mathbb{R})$ is called *Fréchet* $\mathcal{I}$-*differentiable at* $A \in \mathcal{B}_{sa}(\mathcal{H})$ *along* $\mathcal{I}_{sa}$ if

$$\lim_{\|X\|_{\mathcal{I}} \to 0} \frac{1}{\|X\|_{\mathcal{I}}} \left\| f(A + X) - f(A) - D_G^{\mathcal{I}} f(A)(X) \right\|_{\mathcal{I}} = 0$$

for each $X \in \mathcal{I}_{sa}$.

Clearly, Fréchet I-differentiable functions are also Gâteaux I-differentiable. Let Γ be an open set in $\mathbb{R}$ containing 0 and set

$$\mathcal{B}_{sa}(\mathcal{H})(\Gamma) := \{A \in \mathcal{B}_{sa}(\mathcal{H}) : \sigma(A) \subset \Gamma\}.$$

Denote by $\mathrm{GD}_I(\Gamma)$ the space of all Gâteaux I-differentiable functions at all $A \in \mathcal{B}_{sa}(\mathcal{H})(\Gamma)$ along I_{sa} and by $\mathrm{FD}_I(\Gamma)$ the space of all Fréchet I-differentiable functions at all $A \in \mathcal{B}_{sa}(\mathcal{H})(\Gamma)$ along I_{sa}.

I-Lipschitz and Gâteaux I-Differentiable Functions

In many cases the difference between I-Lipschitz and Gâteaux I-differentiable functions is not that big. It is proved in [103, Corollary 3.3] that if $I \neq \mathcal{S}^1$ is separable and an operator $A \in \mathcal{B}_{sa}(\mathcal{H})$ has no eigenvalues, then each I-Lipschitz function g on $\mathbb{R}$ is Gâteaux I-differentiable at A along I_{sa}. In particular, all Lipschitz functions in the usual sense are Gâteaux $\mathcal{S}^p$-differentiable, $1 < p \leqslant \infty$, at all $A \in \mathcal{B}_{sa}(\mathcal{H})$ without eigenvalues along $\mathcal{S}^p_{sa}$. However, for Gâteaux I-differentiability of f at all $A \in \mathcal{B}_{sa}(\mathcal{H})$, we need differentiability of f. The following result is established in [103, Theorem 3.6].

Theorem 5.3.23 *Let Γ be an open set in $\mathbb{R}$ containing 0. Let I be a separable s. n. ideal. Then, the following properties are equivalent.*

(i) $f \subset I\text{-Lip}(\Gamma)$ and f is differentiable on Γ.
(ii) $f \in \mathrm{GD}_I(\Gamma)$ (at all diagonalizable $A \in \mathcal{B}_{sa}(\mathcal{H})(\Gamma)$ if $I = \mathcal{S}^1$).
(iii) f is Gâteaux I-Differentiable at all $A \in \mathcal{B}_{sa}(\mathcal{H})(\Gamma)$ along $\mathfrak{F}_{sa}$ (at all diagonalizable $A \in \mathcal{B}_{sa}(\mathcal{H})(\Gamma)$ if $I = \mathcal{S}^1$).
(iv) f is Gâteaux I-Differentiable at all $A \in \mathfrak{F}_{sa}(\Gamma)$ along $\mathfrak{F}_{sa}$ and, for each compact $I \subset \Gamma$, there is $K_I > 0$ such that $\|D_G^I f(A)\|_I \leqslant K_I$ for all $A \in \mathfrak{F}_{sa}(I)$.

For ideals $\mathcal{S}^p$, $1 < p < \infty$, the result of Theorem 5.3.23 is improved in [110, Theorem 7.15] (a refinement of Theorem 5.3.8): $f \in \mathrm{GD}_{\mathcal{S}^p}(\Gamma)$ if and only if f is differentiable on Γ and has bounded derivative on all compact subsets of Γ. Unlike the spaces $\mathrm{GD}_{\mathcal{S}^p}(\Gamma)$, $1 < p < \infty$, there is no full description of the spaces $\mathrm{GD}_{\mathcal{S}^1}(\Gamma)$, $\mathrm{GD}_{\mathcal{S}^\infty}(\Gamma)$, $\mathrm{GD}_{\mathcal{B}(\mathcal{H})}(\Gamma)$. However, the theorem below links the spaces $\mathrm{GD}_{\mathcal{S}^\infty}(\Gamma)$ and $\mathcal{S}^\infty\text{-Lip}(\Gamma)$. It follows from [103, Theorem 3.7, Corollary 3.10].

Theorem 5.3.24 *Let $\Gamma \subseteq \mathbb{R}$ be an open set. Then,*

$$\mathrm{GD}_{\mathcal{S}^1}(\Gamma) \subseteq \mathrm{GD}_{\mathcal{S}^\infty}(\Gamma) = \mathcal{S}^1\text{-Lip}(\Gamma) = \mathcal{S}^\infty\text{-Lip}(\Gamma) = \mathcal{B}(\mathcal{H})\text{-Lip}(\Gamma).$$

Problem 5.3.25 Let $I \neq \mathcal{S}^p$, $1 \leqslant p \leqslant \infty$, be a separable s. n. ideal and $f \colon \Gamma \to \mathbb{R}$ a continuous function. Does $f \in \mathrm{GD}_I(\Gamma)$ if and only if f has bounded derivative in all compact subsets of Γ?

Fréchet $\mathcal{I}$-Differentiable Functions

Let $\mathcal{I}$ be a separable s. n. ideal or $\mathcal{B}(\mathcal{H})$ and $\Gamma \subseteq \mathbb{R}$ an open set. Similarly to Theorem 5.3.23, a function $f \in C(\mathbb{R})$ is Fréchet $\mathcal{I}$-differentiable at all $A \in \mathcal{B}_{sa}(\mathcal{H})(\Gamma)$ along $\mathcal{I}_{sa}$ if and only if f is Fréchet $\mathcal{I}$-differentiable at all $A \in \mathcal{B}_{sa}(\mathcal{H})(\Gamma)$ along $\mathfrak{F}_{sa}$ (see [103, Theorem 4.1]). It is proved in [110, Theorem 7.17] (a refinement of Theorem 5.3.8) that $FD_{\mathcal{S}^p}(\Gamma) = C^1(\Gamma)$ for $1 < p < \infty$.

We summarize some results on $\mathcal{S}^\infty$- and $\mathcal{B}(\mathcal{H})$-differentiability below.

Theorem 5.3.26 *Let $\Gamma \subseteq \mathbb{R}$ be an open set. Then,*

$$C^2(\mathbb{R}) \subsetneqq FD_{\mathcal{B}(\mathcal{H})}(\Gamma) = FD_{\mathcal{S}^\infty}(\Gamma) = GD_{\mathcal{B}(\mathcal{H})}(\Gamma) \subseteq GD_{\mathcal{S}^\infty}(\Gamma) \cap C^1(\Gamma).$$
$$(5.3.15)$$

Proof By, for instance, [73], $C^2(\mathbb{R}) \subset FD_{\mathcal{B}(\mathcal{H})}(\Gamma)$. Moreover, by [103, Proposition 2.4], $FD_{\mathcal{B}(\mathcal{H})}(\Gamma) \subseteq C^1(\Gamma)$. By [103, Theorem 4.3], $FD_{\mathcal{B}(\mathcal{H})}(\Gamma) = FD_{\mathcal{S}^\infty}(\Gamma) = GD_{\mathcal{B}(\mathcal{H})}(\Gamma)$. The latter implies $GD_{\mathcal{B}(\mathcal{H})}(\Gamma) \subseteq \mathcal{B}(\mathcal{H})\text{-Lip}(\Gamma) \cap C^1(\Gamma)$. Recalling $GD_{\mathcal{S}^\infty}(\Gamma) = \mathcal{B}(\mathcal{H})\text{-Lip}(\Gamma)$ (see Theorem 5.3.24) completes the proof of (5.3.15).

We also recall that $\mathcal{B}(\mathcal{H})\text{-Lip}(\Gamma)$ contains functions not in $C^1(\Gamma)$ (see Theorem 5.1.13) and there are functions in $C^1(\Gamma)$ that do not belong to $\mathcal{B}(\mathcal{H})\text{-Lip}(\Gamma)$ (see Remark after Theorem 5.1.8). □

The relation between the spaces of Gâteaux differentiable and operator Lipschitz functions on I is studied in [103] and summarized in Theorem 5.3.28 below. We need the following auxiliary result due to [103, Proposition 7.2 and Theorem 7.3].

Theorem 5.3.27 *Let I be a compact subset of $\mathbb{R}$. Let $\|\cdot\|_I$ denote the norm on $\mathcal{B}(\mathcal{H})\text{-Lip}(I)$ given by*

$$\|f\|_I := \sup_{t \in I} |f(t)| + k(f, I),$$

where $k(f, I)$ is the minimal constant D satisfying (5.1.14). Then, the following assertions hold.

(i) $(\mathcal{B}(\mathcal{H})\text{-Lip}(I), \|\cdot\|_I)$ is a commutative Banach $*$-algebra.
(ii) If I has a non-empty interior, then $(\mathcal{B}(\mathcal{H})\text{-Lip}(I), \|\cdot\|_I)$ is not separable.

Let Γ be an open set in $\mathbb{R}$. The space $\mathcal{B}(\mathcal{H})\text{-Lip}(\Gamma)$ is a commutative $*$-algebra. Endowed with the family of seminorms $\{\|\cdot\|_I\}_I$, where I are compacts in Γ, the algebra $(\mathcal{B}(\mathcal{H})\text{-Lip}(\Gamma), \{\|\cdot\|_I\}_I)$ becomes a complete space which is not separable by the theorem above. The following result is derived in [103, Theorem 7.9 and Theorem 7.12].

Theorem 5.3.28 *Let Γ be an open set in $\mathbb{R}$. The space $GD_{\mathcal{B}(\mathcal{H})}(\Gamma)$ is a closed subalgebra of $\mathcal{B}(\mathcal{H})\text{-Lip}(\Gamma)$. Moreover, the polynomials are dense in $GD_{\mathcal{B}(\mathcal{H})}(\Gamma)$, so $GD_{\mathcal{B}(\mathcal{H})}(\Gamma)$ is a separable $*$-subalgebra of nonseparable $\mathcal{B}(\mathcal{H})\text{-Lip}(\Gamma)$.*

5.4 Taylor Approximation of Operator Functions

Let $A, B \in \mathcal{D}_{sa}$ and U be unitary. Suppose f is a C^n function on $\mathbb{R}$ or $\mathbb{T}$ as applicable and consider the Taylor remainders

$$R_{n,f,A}(B) := f(A+B) - \sum_{k=0}^{n-1} \frac{1}{k!} \frac{d^k}{dt^k} f(A+tB)\Big|_{t=0} \tag{5.4.1}$$

in the self-adjoint case and

$$Q_{n,f,U}(B) := f(e^{iB}U) - \sum_{k=0}^{n-1} \frac{1}{k!} \frac{d^k}{dt^k} f(e^{itB}U)\Big|_{t=0} \tag{5.4.2}$$

in the unitary case.

The Taylor remainders $R_{1,f,A}(B) = f(A+B) - f(A)$ and $Q_{1,f,U}(B) = f(e^{iB}U) - f(U)$ of order $n = 1$ are studied in Sect. 5.1. In this section we discuss existence and estimates for remainders of order $n \geqslant 2$.

5.4.1 Taylor Remainders of Matrix Functions

The theorem below demonstrates that the study of matrix Taylor remainders can be reduced to the study of multilinear Schur multipliers. In the particular case $n = 2$ this result is justified in [53, Theorem 16] and [54, Theorem 6]. The extension to an arbitrary $n \geqslant 2$ in the self-adjoint case is relatively standard; the extension in the unitary case is proved in [205].

Theorem 5.4.1

(i) *Let f be n times differentiable on $\mathbb{R}$, $A, B \in \mathcal{B}_{sa}(\ell_d^2)$. Then, for the remainder defined in (5.4.1),*

$$R_{n,f,A}(B) = T_{f^{[n]}}^{A+B,A,\ldots,A}(B,\ldots,B). \tag{5.4.3}$$

(ii) *Suppose $f \in C^n(\mathbb{T})$, A is a self-adjoint and U unitary operators in $\mathcal{B}(\ell_d^2)$. Then, for the remainder defined in (5.4.2),*

$$Q_{n,f,U}(A)$$

$$= \sum_{l=1}^{n} \sum_{\substack{j_1,\ldots,j_l \geqslant 1 \\ j_1+\cdots+j_l=n}} T_{f^{[l]}}^{e^{iA}U,U,\ldots,U} \left(\sum_{k=j_1}^{\infty} \frac{(iA)^k}{k!} U, \frac{(iA)^{j_2}}{j_2!} U, \ldots, \frac{(iA)^{j_l}}{j_l!} U \right). \tag{5.4.4}$$

Proof

(i) Using (3.1.8) and Theorem 5.3.2, we obtain

$$R_{n,f,A}(B) = T_{f^{[1]}}^{A+B,A}(B) - \sum_{k=1}^{n-1} T_{f^{[k]}}^{A,A,...,A}(B, \ldots, B).$$

Applying Proposition 4.1.5 with $m = 1$ to the difference $T_{f^{[1]}}^{A+B,A}(B) -$ $T_{f^{[1]}}^{A,A}(B)$ on the right hand-side of the equality above, we obtain

$$R_{n,f,A}(B) = T_{f^{[2]}}^{A+B,A,A}(B, B) - \sum_{k=2}^{n-1} T_{f^{[k]}}^{A,A,...,A}(B, \ldots, B).$$

Applying the representation (4.1.9) $(n-1)$-times yields (5.4.3).

(ii) We will prove that for all $1 \leqslant m \leqslant n$ the formula

$$Q_{m,f,U}(A)$$

$$= \sum_{l=1}^{m} \sum_{\substack{j_1,\ldots,j_l \geqslant 1 \\ j_1+\cdots+j_l=m}} T_{f^{[l]}}^{e^{iA}U,U,...,U} \left(\sum_{k=j_1}^{\infty} \frac{(iA)^k}{k!} U, \frac{(iA)^{j_2}}{j_2!} U, \ldots, \frac{(iA)^{j_l}}{j_l!} U \right)$$

$$(5.4.5)$$

holds, using induction. The formula (5.4.4) will follow from (5.4.5) for $m = n$. For $m = 1$ by (3.1.10), we have

$$Q_{1,f,U}(A) = f(e^{iA}U) - f(U) = T_{f^{[1]}}^{e^{iA}U,U}(e^{iA}U - U)$$

$$= T_{f^{[1]}}^{e^{iA}U,U} \left(\sum_{k=1}^{\infty} \frac{(iA)^k}{k!} U \right),$$

that is, (5.4.5) holds for $m = 1$. Assume that (5.4.5) holds for $m = p < n$, that is, we have

$$Q_{p+1,f,U}(A)$$

$$= \sum_{l=1}^{p} \sum_{\substack{j_1,\ldots,j_l \geqslant 1 \\ j_1+\cdots+j_l=p}} T_{f^{[l]}}^{e^{iA}U,U,...,U} \left(\sum_{k=j_1}^{\infty} \frac{(iA)^k}{k!} U, \frac{(iA)^{j_2}}{j_2!} U, \ldots, \frac{(iA)^{j_l}}{j_l!} U \right)$$

$$- \frac{1}{p!} \frac{d^p}{dt^p} f(e^{itA}U) \big|_{t=0}.$$

Applying Theorem 5.3.4 and decomposing $i^p = i^{j_1} \ldots i^{j_l}$, we obtain

$$Q_{p+1,f,U}(A) = \sum_{l=1}^{p} \sum_{\substack{j_1,\ldots,j_l \geqslant 1 \\ j_1+\cdots+j_l=p}} T_{f^{[l]}}^{e^{iA}U,U,\ldots,U}\left(\sum_{k=j_1}^{\infty} \frac{(iA)^k}{k!}U, \frac{(iA)^{j_2}}{j_2!}U,\ldots,\frac{(iA)^{j_l}}{j_l!}U\right)$$

$$- \sum_{l=1}^{p} \sum_{\substack{j_1,\ldots,j_l \geqslant 1 \\ j_1+\cdots+j_l=p}} T_{f^{[l]}}^{U,\ldots,U}\left(\frac{(iA)^{j_1}}{j_1!}U,\ldots,\frac{(iA)^{j_l}}{j_l!}U\right).$$

The latter can be rewritten as

$$Q_{p+1,f,U}(A)$$

$$= \sum_{l=1}^{p} \sum_{\substack{j_1,\ldots,j_l \geqslant 1 \\ j_1+\cdots+j_l=p}} \left(T_{f^{[l]}}^{e^{iA}U,U,\ldots,U}\left(\sum_{k=j_1}^{\infty} \frac{(iA)^k}{k!}U, \frac{(iA)^{j_2}}{j_2!}U,\ldots,\frac{(iA)^{j_l}}{j_l!}U\right)\right.$$

$$- T_{f^{[l]}}^{e^{iA}U,\ldots,U}\left(\frac{(iA)^{j_1}}{j_1!}U,\ldots,\frac{(iA)^{j_l}}{j_l!}U\right)$$

$$\left.+ T_{f^{[l]}}^{e^{iA}U,\ldots,U}\left(\frac{(iA)^{j_1}}{j_1!}U,\ldots,\frac{(iA)^{j_l}}{j_l!}U\right) - T_{f^{[l]}}^{U,\ldots,U}\left(\frac{(iA)^{j_1}}{j_1!}U,\ldots,\frac{(iA)^{j_l}}{j_l!}U\right)\right).$$

Applying multilinearity of the multiple operator integral and the property (4.1.9), we obtain

$$Q_{p+1,f,U}(A)$$

$$= \sum_{l=1}^{p} \sum_{\substack{j_1,\ldots,j_l \geqslant 1 \\ j_1+\cdots+j_l=p}} T_{f^{[l]}}^{e^{iA}U,U,\ldots,U}\left(\sum_{k=j_1+1}^{\infty} \frac{(iA)^k}{k!}U, \frac{(iA)^{j_2}}{j_2!}U,\ldots,\frac{(iA)^{j_l}}{j_l!}U\right)$$

$$+ \sum_{l=1}^{p} \sum_{\substack{j_1,\ldots,j_l \geqslant 1 \\ j_1+\cdots+j_l=p}} T_{f^{[l+1]}}^{e^{iA}U,U,\ldots,U}\left(\sum_{k=1}^{\infty} \frac{(iA)^k}{k!}U, \frac{(iA)^{j_1}}{j_1!}U,\ldots,\frac{(iA)^{j_l}}{j_l!}U\right)$$

$$=: S_1 + S_2. \tag{5.4.6}$$

Making the substitution $i_1 = j_1 + 1$, $i_r = j_r$ for $2 \leqslant r \leqslant l$ in the first sum on the right-hand side of (5.4.6), we obtain

$$S_1 = \sum_{l=1}^{p} \sum_{\substack{i_1 \geqslant 2, i_2,\ldots,i_l \geqslant 1 \\ i_1+\cdots+i_l=p+1}} T_{f^{[l]}}^{e^{iA}U,U,\ldots,U}\left(\sum_{k=i_1}^{\infty} \frac{(iA)^k}{k!}U, \frac{(iA)^{i_2}}{i_2!}U,\ldots,\frac{(iA)^{i_l}}{i_l!}U\right).$$

Relabeling the summands of S_2 by the mapping $l \mapsto l - 1$, and then making the substitution $i_1 = 1$, $i_r = j_{r-1}$ for $2 \leqslant r \leqslant l$, we obtain

$$S_2 = \sum_{\substack{l=2 \\ }}^{p+1} \sum_{\substack{j_1,\dots,j_{l-1} \geqslant 1 \\ j_1+\cdots+j_{l-1}=p}} T_{f^{[l]}}^{e^{iA}U,U,\dots,U}\left(\sum_{k=1}^{\infty} \frac{(iA)^k}{k!}U, \frac{(iA)^{j_1}}{j_1!}U, \dots, \frac{(iA)^{j_{l-1}}}{j_{l-1}!}U\right)$$

$$= \sum_{\substack{l=2 \\ }}^{p+1} \sum_{\substack{i_1=1,i_2,\dots,i_l \geqslant 1 \\ i_1+\cdots+i_l=p+1}} T_{f^{[l]}}^{e^{iA}U,U,\dots,U}\left(\sum_{k=i_1}^{\infty} \frac{(iA)^k}{k!}U, \frac{(iA)^{i_2}}{i_2!}U, \dots, \frac{(iA)^{i_l}}{i_l!}U\right).$$

Combining the two previous equalities with (5.4.6), we arrive at

$$Q_{p+1,f,U}(A) = T_{f^{[1]}}^{e^{iA}U,U}\left(\sum_{k=p+1}^{\infty} \frac{(iA)^k}{k!}\right)$$

$$+ \sum_{l=2}^{p}\left(\sum_{\substack{i_1=1,i_2,\dots,i_l \geqslant 1 \\ i_1+\cdots+i_l=p+1}} + \sum_{\substack{i_1 \geqslant 2,i_2,\dots,i_l \geqslant 1 \\ i_1+\cdots+i_l=p+1}}\right)$$

$$\times T_{f^{[l]}}^{e^{iA}U,U,\dots,U}\left(\sum_{k=i_1}^{\infty} \frac{(iA)^k}{k!}U, \frac{(iA)^{i_2}}{i_2!}U, \dots, \frac{(iA)^{i_l}}{i_l!}U\right)$$

$$+ T_{f^{[p+1]}}^{e^{iA}U,U,\dots,U}\left(\sum_{k=1}^{\infty} \frac{(iA)^k}{k!}U, iAU, \dots, iAU\right)$$

$$= \sum_{\substack{l=1 \\ }}^{p+1} \sum_{\substack{i_1,i_2,\dots,i_l \geqslant 1 \\ i_1+\cdots+i_l=p+1}} T_{f^{[l]}}^{e^{iA}U,U,\dots,U}\left(\sum_{k=i_1}^{\infty} \frac{(iA)^k}{k!}U, \frac{(iA)^{i_2}}{i_2!}U, \dots, \frac{(iA)^{i_l}}{i_l!}U\right),$$

which proves (5.4.5).   □

A coarse Hilbert-Schmidt bound for Taylor remainders of matrix functions can be found in [88, Theorem 2.3.1 (4)]. More delicate estimates for Taylor remainders can be derived from estimates for multiple operator integrals discussed in this book. In particular, the following estimate holds.

Theorem 5.4.2 *Let $A, B \in \mathcal{B}_{sa}(\ell_d^2)$ and $[a, b]$ be the convex hull containing $\sigma(A) \cup \sigma(A + B)$. Let $1 < p < \infty$, $n \in \mathbb{N}$, and let f be n times differentiable on $[a, b]$ such that $f^{(n)}$ is bounded. Then, there exists $c_{p,n} > 0$ such that*

$$\|R_{n,f,A}(B)\|_p \leqslant c_{p,n}\|f^{(n)}\|_{L^{\infty}[a,b]}\|B\|_{pn}^n.$$

The result of Theorem 5.4.2 is extended to the general infinite-dimensional case in the next subsection, where the bounds in the unitary case are also discussed.

5.4.2 Taylor Remainders for Perturbations in S^p and $\mathcal{B}(\mathcal{H})$

In this subsection we collect estimates for Taylor remainders of operator functions with respect to different norms. As in the finite dimensional case, the study of operator Taylor remainders can be reduced to the study of multilinear operator integrals.

Self-adjoint Case

Theorem 5.4.3 *Let $A \in \mathcal{D}_{sa}$, $B \in \mathcal{B}_{sa}(\mathcal{H})$, $n \in \mathbb{N}$, and $f \in B^1_{\infty 1}(\mathbb{R}) \cap B^n_{\infty 1}(\mathbb{R})$. Then, the Taylor remainder admits the representation*

$$R_{n,f,A}(B) = \frac{1}{(n-1)!} \int_0^1 (1-t)^{n-1} \frac{d^n}{ds^n} f(A+sB)\big|_{s=t} \, dt, \qquad (5.4.7)$$

where the integral converges in the strong operator topology. Moreover, if $B \in S^p$, $n \leqslant p < \infty$, then the integral converges in $S^{p/n}$-norm and

$$R_{n,f,A}(B) \in S^{p/n}. \qquad (5.4.8)$$

Proof It follows from (5.3.9) and Proposition 4.3.15(i) that the derivative $t \rightarrow \frac{d^n}{ds^n} f(A+sB)\big|_{s=t}$ is continuous on $[0, 1]$ in the strong operator topology and in $S^{p/n}$. Hence, we have the integral representation for the remainder (5.4.7), which can be proved as in [181, Theorem 1.43] by applying functionals in the dual space $(\mathcal{B}(\mathcal{H}))^*$ and reducing the problem to the scalar case. Since $t \rightarrow \frac{d^n}{ds^n} f(A+sB)\big|_{s=t}$ is continuous in $S^{p/n}$, (5.4.8) holds. $\square$

Combining the estimate (5.3.10) and representation (5.4.7) implies the following estimate for the remainder.

Theorem 5.4.4 *Let $A \in \mathcal{D}_{sa}$, $B \in \mathcal{B}_{sa}(\mathcal{H})$, $n \in \mathbb{N}$, $f \in B^1_{\infty 1}(\mathbb{R}) \cap B^n_{\infty 1}(\mathbb{R})$, and $1 \leqslant p \leqslant \infty$. Then,*

$$\big\| R_{n,f,A}(B) \big\|_p \leqslant c_n \|f\|_{B^n_{\infty 1}(\mathbb{R})} \|B\|_{pn}^n. \qquad (5.4.9)$$

The estimate (5.4.9) can be improved and the set of functions f for which it holds can be substantially enlarged when $1 < p < \infty$. The next result for operator Taylor remainders centered at a bounded self-adjoint operator A is obtained in [169, Theorem 4.1].

Theorem 5.4.5 *Let* $1 < p < \infty$, $n \in \mathbb{N}$. *Then, there exists* $c_{n,p} > 0$ *such that for* $f \in C^n(\mathbb{R})$, $A = A^* \in \mathcal{B}(\mathcal{H})$, *and* $B = B^* \in \mathcal{S}^p$,

$$\|R_{n,f,A}(B)\|_p \leqslant c_{p,n}\|f^{(n)}\|_\infty \|B\|_{pn}^n,$$

where $R_{n,f,A}(B)$ *is defined in* (5.4.1).

Proof Since we work with bounded operators, we assume without loss of generality that f is supported in some compact set containing the spectra of operators, that is, $f \in C_c^n(\mathbb{R})$. Let P_r, $0 \leqslant r < 1$, be the Poisson kernel. For $k \in \mathbb{N}$ define the convolution $f_k := P_{1/k} * f$. Then,

$$\|f_k^{(m)} - f^{(m)}\|_\infty \to 0, \quad k \to \infty,$$

for all $0 \leqslant m \leqslant n$, since $f_k^{(m)} = P_{1/k} * f^{(m)}$. Moreover, $\{f_k\}_{k=1}^\infty \subset W_n(\mathbb{R})$, (see, e.g., [198]).

By (5.4.7) and (5.3.12) we obtain

$$\|R_{n,f_k,A}(B)\|_p \leqslant c_{n,p}\|f_k^{(n)}\|_\infty \|B\|_{pn}^n.$$

Again using $f_k^{(n)} = P_{1/k} * f^{(n)}$ and Young's inequality, since $\|P_{1/k}\|_1 \leqslant 1$, for all $k \in \mathbb{N}$, we obtain

$$\|f_k^{(n)}\|_\infty \leqslant \|P_{1/k}\|_1 \|f^{(n)}\|_\infty \leqslant \|f^{(n)}\|_\infty$$

and so

$$\|R_{n,f_k,A}(B)\|_p \leqslant c_{n,p}\|f^{(n)}\|_\infty \|B\|_{pn}^n. \tag{5.4.10}$$

We now prove that

$$\mathrm{Tr}\big(R_{n,f_k,A}(B)C\big) \to \mathrm{Tr}\big(R_{n,f,A}(B)C\big) \quad \text{as } k \to \infty \tag{5.4.11}$$

for every $C \in \mathcal{S}^1$. Indeed,

$$\|f_k(A + B) - f(A + B)\| \leqslant \|f_k - f\|_\infty \to 0$$

and, similarly,

$$\|f_k(A) - f(A)\| \leqslant \|f_k - f\|_\infty \to 0.$$

Moreover, for all $1 \leqslant q \leqslant n - 1$, using subsequently the representation (5.3.9), Theorem 4.3.11, and the fact that $f \in \cap_{m=1}^{n-1} W_m(\mathbb{R})$, we obtain

$$\left\| \frac{d^q}{dt^q} f_k(A + tB) \big|_{t=0} - \frac{d^q}{dt^q} f(A + tB) \big|_{t=0} \right\|_p$$

$$= \left\| T^{A,\ldots,A}_{f_k^{[q]}}(B, \ldots, B) - T^{A,\ldots,A}_{f^{[q]}}(B, \ldots, B) \right\|_p$$

$$= \left\| T^{A,\ldots,A}_{(f_k - f)^{[q]}}(B, \ldots, B) \right\|_p$$

$$\leqslant c_{p,q} \left\| f_k^{(q)} - f^{(q)} \right\|_\infty \|B\|_{pq}^q \to 0, \quad k \to \infty.$$

Along with Hölder's inequality, the latter implies (5.4.11) for every $C \in S^1 \subset (S^p)^*$. Finally, by the properties (5.4.10), (5.4.11), Hölder's inequality, and denseness of S^1 in $(S^p)^* = S^{p/(p-1)}$, we conclude

$$\|R_{n,f,A}(B)\|_p = \sup_{C \in S^1, \, \|C\|_{p/(p-1)} \leqslant 1} \left| \mathrm{Tr}\left(R_{n,f_k,A}(B)C \right) \right| \leqslant c_{n,p} \cdot \|f^{(n)}\|_\infty \cdot \|B\|_{pn}^n.$$

As a consequence of Theorem 5.3.10 along with extension of Theorem 4.3.13 to functions that are differentiable, but not continuously, and a version of Theorem 4.3.10 for the transformation Γ given by Definition 4.2.1 that are derived in [52], we obtain the following estimate for operator Taylor remainders centered at an unbounded self-adjoint operator A.

Theorem 5.4.6 *Let* $1 < p < \infty$, $n \in \mathbb{N}$ *and* $A \in \mathcal{D}_{sa}$. *Let* f *be* n *times differentiable on* $\mathbb{R}$ *such that* $f', \ldots, f^{(n)}$ *are bounded and let* $X = X^* \in S^{pn}$. *Then,*

$$\left\| f(A + X) - f(A) - \sum_{k=1}^{n-1} \frac{1}{k!} D_{G,p}^k f(A)(X, \ldots, X) \right\|_p \leqslant c_{p,n} \|f^{(n)}\|_\infty \|X\|_{pn}^n.$$

When $p = 1$, the condition $f \in C^n(\mathbb{R})$ is not sufficient for the remainder to be in S^1. A counterexample with unbounded A in the case $n = 2$ is constructed in [53]. The following result for a bounded A and general n is derived in [169, Theorem 5.1].

Theorem 5.4.7 *Let* $n \in \mathbb{N}$. *There exist a separable Hilbert space* $\mathcal{H}$ *and self-adjoint operators* $A \in \mathcal{B}(\mathcal{H})$ *and* $B \in S^n(\mathcal{H})$ *such that*

$$R_{n,f_n,A}(B) \notin S^1(\mathcal{H}),$$

for the function $f_n \in C^n(\mathbb{R})$ *given by*

$$f_n(x) := x^{n-1} \begin{cases} |x| \left(\log |\log |x|| - 1 \right)^{-\frac{1}{2}}, & x \neq 0 \\ 0, & x = 0. \end{cases} \tag{5.4.12}$$

The proof of Theorem 5.4.7 builds on the following dimension dependent bound due to [169, Theorem 5.5].

Theorem 5.4.8 *Let* $n, d \in \mathbb{N}$, $d \geqslant 2$. *For* A_d, $B_d \in \mathcal{B}_{sa}(\ell^2_{2d})$ *satisfying Theorem 5.1.11, there are operators* $X_1, \ldots X_n \in \mathcal{B}(\ell^2_{2d})$ *with* $\|X_j\|_n = 1$, $1 \leqslant j \leqslant n \in \mathbb{N}$, *such that*

$$\left\| T^{A_d+B_d, A_d, \ldots, A_d}_{f^{[n]}_n}(X_1, \ldots, X_n) \right\|_1 \geqslant \text{const} \, (\log d)^{\frac{1}{2}}$$

for a function $f_n : \mathbb{R}^{n+1} \to \mathbb{R}$ *given by* (5.4.12).

Proof By (3.1.8) and Theorem 5.1.11,

$$\left\| T^{A_d+B_d, A_d}_{h^{[1]}}(B_d) \right\| = \| h(A_d + B_d) - h(A_d) \| \geqslant \text{const} \, (\log d)^{\frac{1}{2}} \|B_d\|_\infty.$$

Hence, by Theorem 4.1.9((iii)),

$$\left\| T^{A_d+B_d, A_d}_{h^{[1]}} : \mathcal{S}^1_{2d} \to \mathcal{S}^1_{2d} \right\| = \left\| T^{A_d+B_d, A_d}_{h^{[1]}} : \mathcal{B}(\ell^2_{2d}) \to \mathcal{B}(\ell^2_{2d}) \right\| \geqslant \text{const} \, (\log d)^{\frac{1}{2}}.$$

It is easy to check that

$$f^{[n]}_n(x_0, 0 \ldots, 0, x_n) = h^{[1]}(x_0, x_n).$$

Combining the latter with Theorem 4.1.9(i) (where $p_1 = \cdots = p_n = n$ and $p = 1$) implies

$$\left\| T^{A_d+B_d, A_d, \ldots, A_d}_{f^{[n]}_n} : \mathcal{S}^n_{2d} \times \cdots \times \mathcal{S}^n_{2d} \to \mathcal{S}^1_{2d} \right\| \geqslant \left\| T^{A_d+B_d, A_d}_{h^{[1]}} : \mathcal{S}^1_{2d} \to \mathcal{S}^1_{2d} \right\|$$

$$\geqslant \text{const} \, (\log d)^{\frac{1}{2}},$$

completing the proof.   □

Unitary Case

The following result for functions of unitary operators is established in [169, Theorem 4.2].

Theorem 5.4.9 *Let* $1 < p < \infty$, $n \in \mathbb{N}$. *Then, there exists* $c_{n,p} > 0$ *such that for* $\varphi \in C^n(\mathbb{T})$, $A = A^* \in \mathcal{S}^p$, *and a unitary* $U \in \mathcal{B}(\mathcal{H})$,

$$\| Q_{n,\varphi,U}(A) \|_p \leqslant c_{p,n} \sum_{l=1}^{n} \| \varphi^{(l)} \|_\infty \|A\|^n_{pn},$$

where $Q_{n,\varphi,U}(A)$ *is defined in* (5.4.2).

The aforementioned result cannot be extended to the case $p = 1$. A counterexample for $n = 2$ is obtained in [54] and in the case of a general n in [169, Theorem 5.11].

Theorem 5.4.10 *Let $n \in \mathbb{N}$. There exist a separable Hilbert space $\mathcal{H}$, a unitary operator $U \in \mathcal{B}(\mathcal{H})$, and a self-adjoint operator $B \in S^n(\mathcal{H})$ such that*

$$Q_{n,\varphi_n,U}(B) \notin S^1(\mathcal{H}),$$

for the function $\varphi_n \in C^n(\mathbb{T})$ given by

$$\varphi_n(z) := (z - 1)^{n-1}u(z), \tag{5.4.13}$$

where u is given by (5.1.9).

The proof of Theorem 5.4.10 is derived from the dimension-dependent bound of the next consequence of Theorem 5.1.17 obtained in [169, Theorem 5.15].

Theorem 5.4.11 *Let $d \in \mathbb{N}$, $d \geqslant 3$. Then, there exist unitary operators $H_d, K_d \in \mathcal{B}(\ell^2_{2d+1})$ satisfying Theorem 5.1.17 and such that*

$$\left\| T^{K_d,H_d,\dots,H_d}_{\varsigma} : S^n_{2d+1} \times \cdots \times S^n_{2d+1} \to S^1_{2d+1} \right\| \geqslant \mathrm{const}\,(\log d)^{\frac{1}{2}}$$

for $\varsigma : \mathbb{T}^{n+1} \to \mathbb{C}$ given by

$$\varsigma(z_0, \dots, z_n) := z_1 \cdot z_2 \cdot \cdots \cdot z_{n-1} \cdot \varphi_n^{[n]}(z_0, \dots, z_n)$$

with φ_n defined in (5.4.13).

5.4.3 Taylor Remainders for Unsummable Perturbations

Unsummable perturbations naturally arise in the study of differential operators, as discussed in Sect. 5.5.1. Summability of the respective approximation remainders in such cases is ensured by suitable summability restrictions on resolvents of operators as well as by choice of sufficiently nice scalar functions. When unsummable perturbations produce unsummable Taylor remainders, non-Taylor approximations can be used, as demonstrated in Sect. 5.5.1.

Let $(\mathcal{M}, \tau)$ be a semifinite von Neumann algebra equipped with a normal faithful semifinite trace. It follows from [23, Lemmas 1.4 and 1.7] that if $A\eta\mathcal{M}_{sa}$ and A has τ-compact resolvent, then $f(A) \in \mathcal{L}^1(\mathcal{M}, \tau)$ for every $f \in C_c(\mathbb{R})$. More generally, it is proved in [193, Theorem 3.1] that multiple operator integrals built over H, $f^{[n]}$ and applied to tuples of operators in $\mathcal{M}$ attain their values in $\mathcal{L}^1(\mathcal{M}, \tau)$ for $f \in C_c^{n+1}(\mathbb{R})$. A consequence of this result and a specific estimate for the operator Taylor remainder are established in [193, Theorem 4.1] stated below.

Theorem 5.4.12 *Let $A \eta \mathcal{M}_{sa}$ have τ-compact resolvent, $B \in \mathcal{M}_{sa}$, $n \in \mathbb{N}$, and $\epsilon > 0$. Then, for $f \in C_c^{n+1}((a, b))$, $R_{n,f,A}(B) \in \mathcal{L}^1(\mathcal{M}, \tau)$ and*

$$|\tau(R_{n,f,A}(B))| \leqslant \|f^{(n)}\|_\infty \|B\|^n \left(2^n(n+1) + c_n\right)(b - a + 1)^n \left(1 + \Omega_{a,b,A,B}\right)$$

$$\times \left(\Omega_{a_\epsilon, b_\epsilon, A, B} + \sqrt{2}\,(b_\epsilon - a_\epsilon + 1)^{\frac{3}{2}} \max_{1 \leqslant k \leqslant n} \frac{\|\varphi_\epsilon^{(k+1)}\|_\infty}{k!}\right),$$

where $c_n > 0$, $\Omega_{a,b,A,B}$ is such that

$$\sup_{t \in [0,1]} \tau\left(E_{A+tB}((a, b))\right) \leqslant \Omega_{a,b,A,B}$$

and

$$\mu_{\Omega_{a,b,A,B}}\left((I + A^2)^{-1}\right) \leqslant \frac{1}{(1 + \max\{a^2, b^2\})(1 + \|B\| + \|B\|^2)}$$

hold and φ_ϵ is a smoothening of the indicator function of (a, b) satisfying $\sqrt[4]{\varphi_\epsilon} \in C_c^\infty((a_\epsilon, b_\epsilon))$, $\varphi_\epsilon|_{(a,b)} \equiv 1$, and $0 \leqslant \varphi_\epsilon \leqslant 1$, where $a_\epsilon = a - \epsilon$, $b_\epsilon = b + \epsilon$.

Theorem 5.4.12 is proved by creating weights that reduce the problem to the case of summable perturbations.

5.5   Spectral Shift

A spectral shift function originates from I. M. Lifshits' work on quantum theory of crystals summarized in [120] that was followed by M. G. Krein's seminal paper [113] starting a mathematical theory of this object. The spectral shift function has evolved into a fundamental object in perturbation theory. It enjoyed several breakthroughs in recent years thanks to development of a new approach to multiple operator integration. There are several surveys dedicated to theory of spectral shift functions of general operators and many papers on spectral shift functions for specific models. The earlier development of the subject and applications can be found in [42, 183, 214], its development up to 2013 is surveyed in [188]. Connections of the spectral shift function to such important objects of perturbation theory as scattering phase and perturbation determinant are briefly surveyed in [33]; connection of the spectral shift function to a spectral flow is discussed in Sect. 5.6. In this subsection, we state major results on the spectral shift function for general operators, emphasizing results obtained beginning 2014.

5.5.1 Spectral Shift Function for Self-adjoint Operators

Summable Perturbations

The following fundamental result is established in [111, 113], and [163, Theorem 1.1] in the cases $n = 1$, $n = 2$, and $n \geqslant 3$, respectively.

Theorem 5.5.1 *If $H \in \mathcal{D}_{sa}$ and $V \in \mathcal{B}_{sa}(\mathcal{H})$ are such that*

$$V \in \mathcal{S}^n, \tag{5.5.1}$$

$n \in \mathbb{N}$, then there exist a unique real-valued function $\eta_n = \eta_{n,H,V} \in L^1(\mathbb{R})$, called the n-th order spectral shift function associated with the pair of operators H and $H + V$, and a constant $c_n > 0$ such that

$$\|\eta_n\|_1 \leqslant c_n \|V\|_n^n$$

and

$$\mathrm{Tr}\left(f(H+V) - \sum_{k=0}^{n-1} \frac{1}{k!} \frac{d^k}{ds^k} f(H+sV)\Big|_{s=0}\right) = \int_{\mathbb{R}} f^{(n)}(t)\, \eta_n(t)\, dt, \tag{5.5.2}$$

for $f \in W_n(\mathbb{R})$.

Remark 5.5.2

(i) The condition (5.5.1) arises in the study of perturbations of discrete Laplacians on a lattice.

(ii) The formula (5.5.2) is extended from $W_n(\mathbb{R})$ to the Besov class $B_{\infty 1}^n(\mathbb{R}) \cap B_{\infty 1}^1(\mathbb{R})$ in [143, 145], and [7] for $n = 1$, $n = 2$, and $n \geqslant 3$, respectively. It is proved in [148] that (5.5.2) with $n = 1$ holds if and only if f is operator Lipschitz, which solves the problem posed in [113], and in [55] that (5.5.2) with $n = 2$ holds for f for which the divided difference $f^{[2]}$ admits a certain Hilbert space factorization. We recall that differentiability of operator functions is discussed in Sect. 5.3. The trace on the left hand side of (5.5.2) is well defined by (5.4.8).

(iii) When $\mathcal{H} = \ell_d^2$, the formula (5.5.2) extends to $f \in C^n[a, b]$, where $[a, b]$ is the convex hull of the set $\sigma(H) \cup \sigma(H + V)$ [194, Theorem 2.1].

The original Krein's proof of (5.5.2) for $n = 1$ is complex analytic in essence. The result is derived explicitly for rank one V, then continued to the case of finite rank V, and by approximations induced for an arbitrary $V \in \mathcal{S}^1$. A purely real analytic proof is constructed in [165] via several stages of approximations. Existence of η_n, $n \geqslant 2$, is established implicitly apart from some particular cases.

Proof (Proof of Theorem 5.5.1 for $n \geqslant 2$) Evaluating the trace in (5.4.7) gives

$$\mathrm{Tr}\big(R_{n,f,H}(V)\big) = \frac{1}{(n-1)!} \int_0^1 (1-t)^{n-1} \mathrm{Tr}\left(\frac{d^n}{ds^n} f(H+sV)\big|_{s=t}\right) dt,$$

where the remainder $R_{n,f,H}(V)$ is given by (5.4.1). By (5.3.13), there exists $c_n > 0$ such that

$$|\mathrm{Tr}(R_{n,f,H}(V))| \leqslant c_n \|V\|_n^n \|f^{(n)}\|_\infty. \tag{5.5.3}$$

By the Riesz representation theorem for functionals in $(C_c^{n+1}(\mathbb{R}))^*$, there exists a Borel measure ν_n such that

$$\|\nu_n\| \leqslant c_n \|V\|_n^n \tag{5.5.4}$$

and

$$\mathrm{Tr}(R_{n,f,H}(V)) = \int_{\mathbb{R}} f^{(n)}(t) \, d\nu_n(t), \tag{5.5.5}$$

for every $f \in C_c^{n+1}(\mathbb{R})$. By approximations, (5.5.5) extends to the space $W_n(\mathbb{R})$.

We prove below that the measure ν_n in (5.5.5), $n \geqslant 2$, is absolutely continuous.

Assume first that $V \in \mathcal{S}^1$. Then, for every $f \in C_c^n(\mathbb{R})$, integration by parts in (5.5.5) with $n-1$ gives

$$\mathrm{Tr}(R_{n-1,f,H}(V)) = - \int_{\mathbb{R}} f^{(n)}(t)\nu_{n-1}((-\infty,t)) \, dt. \tag{5.5.6}$$

By (5.3.13) and the Riesz representation theorem, there exists a Borel measure μ_{n-1} such that

$$\|\mu_{n-1}\| \leqslant c_{n-1} \|V\|_{n-1}^{n-1}$$

and

$$\mathrm{Tr}\left(\frac{d^{n-1}}{ds^{n-1}} f(H+sV)\big|_{s=0}\right) = \int_{\mathbb{R}} f^{(n-1)}(t) \, d\mu_{n-1}(t)$$

$$= - \int_{\mathbb{R}} f^{(n)}(t)\mu_{n-1}((-\infty,t)) \, dt, \tag{5.5.7}$$

for every $f \in C_c^n(\mathbb{R})$. Combining (5.5.6) and (5.5.7) implies

$$\mathrm{Tr}(R_{n,f,H}(V)) = \int_{\mathbb{R}} f^{(n)}(t) \big(\mu_{n-1}((-\infty,t)) - \nu_{n-1}((-\infty,t))\big) \, dt.$$

Thus, (5.5.2) in the case $V \in \mathcal{S}^1$ holds with

$$\eta_n(t) = \mu_{n-1}((-\infty, t)) - \nu_{n-1}((-\infty, t)).$$

It follows from (5.5.4) that

$$\|\eta_n\|_1 \leqslant c_n \|V\|_n^n. \tag{5.5.8}$$

Let $V \in \mathcal{S}^n$ and let $\{V_k\}_{k=1}^{\infty} \subset \mathcal{S}^1$ be such that $\lim_{k\to\infty} \|V - V_k\|_n = 0$. Consider the sequence $\{\eta_{n,H,V_k}\}_{k=1}^{\infty}$ satisfying

$$\mathrm{Tr}(R_{n,f,H}(V_k)) = \int_{\mathbb{R}} f^{(n)}(t)\, \eta_{n,H,V_k}(t)\, dt, \quad k \in \mathbb{N}. \tag{5.5.9}$$

By duality, we obtain

$$\int_{\mathbb{R}} \left| \eta_{n,H,V_j}(t) - \eta_{n,H,V_k}(t) \right| dt$$

$$= \sup_{f \in C_c^{n+1},\, \|f^{(n)}\|_\infty \leqslant 1} \left| \int_{\mathbb{R}} \left(\eta_{n,H,V_j}(t) - \eta_{n,H,V_k}(t) \right) f^{(n)}(t)\, dt \right|. \tag{5.5.10}$$

By the result of the previous paragraph,

$$\left| \int_{\mathbb{R}} \left(\eta_{n,H,V_j}(t) - \eta_{n,H,V_k}(t) \right) f^{(n)}(t)\, dt \right| = \left| \mathrm{Tr}(R_{n,f,H}(V_j) - R_{n,f,H}(V_k)) \right|. \tag{5.5.11}$$

By the uniform continuity of $V \mapsto R_{n,f,H}(V)$, which follows from (5.5.3), and the representations (5.5.10) and (5.5.11),

$$\lim_{j,k\to\infty} \int_{\mathbb{R}} \left| \eta_{n,H,V_j}(t) - \eta_{n,H,V_k}(t) \right| dt = 0.$$

Thus, the sequence $\{\eta_{n,H,V_k}\}_{k=1}^{\infty}$ converges to an integrable function, which we denote by $\eta_{n,H,V}$. Applying (5.5.8) to η_{n,H,V_k} for every $k \in \mathbb{N}$, we deduce the bound

$$\|\eta_{n,H,V}\|_1 \leqslant c_n \|V\|_n^n.$$

Passing to the limit in (5.5.9) as $k \to \infty$ completes the proof of the theorem. $\qquad \square$

The name "spectral shift function" was given to η_1 by M. G. Krein. A reason for this name can be understood from the formula

$$\eta_1(\lambda) = \text{Tr}\big(E_H((-\infty, \lambda))\big) - \text{Tr}\big(E_{H+V}((-\infty, \lambda))\big)$$

$$= \text{card}\{\lambda \in \sigma(H) : \ \lambda < t\} - \text{card}\{\lambda \in \sigma(H+V) : \ \lambda < t\}$$

holding for H and V finite matrices. For $n \geqslant 2$, not only shift of eigenvalues, but also displacement of eigenvectors comes into play. The spectral shift nature of η_n in the case of commuting H and V is demonstrated in [187]. In the case of noncommuting H and V, the larger the value of n, the more intricate η_n is. The following representation for η_n is established in the case $V \in \mathcal{S}^2$ for bounded H in [74, Theorem 5.1 (ii)] and for an unbounded H in [186, Theorem 4.1].

Theorem 5.5.3 *Let $H \in \mathcal{D}_{sa}$, let $V = V^* \in \mathcal{S}^2$, and let $n \in \mathbb{N}$, $n \geqslant 2$. Then,*

$$\eta_n(t) = \frac{\text{Tr}(V^{n-1})}{(n-1)!} - \int_{-\infty}^{t} \eta_{n-1}(s)\, ds \tag{5.5.12}$$

$$- \frac{1}{(n-1)!}$$

$$\times \int_{\mathbb{R}^{n-1}} \big((\lambda - t)_+^{n-2}\big)^{[n-2]}(\lambda_1, \dots, \lambda_{n-1})\, d\,\text{Tr}\big(E_H(\lambda_1)V \dots E_H(\lambda_{n-1})V\big),$$

for a.e. $t \in \mathbb{R}$.

The function x_+^k appearing in (5.5.12) is defined by

$$x_+^k := \begin{cases} x^k & \text{if } x > 0 \\ 0 & \text{if } x \leqslant 0 \end{cases}, \quad \text{for } k \in \mathbb{N} \cup \{0\},$$

where we use the convention $x^0 = 1$ for $x > 0$. For the divided difference appearing in (5.5.12) we use the convention $\big((\lambda - t)_+^{n-2}\big)^{[n-2]}(\lambda_1, \dots, \lambda_{n-1}) = 0$ if $\lambda_1 = \dots = \lambda_{n-1} = t$.

We note that the representation (5.5.12) does not hold for $V \in \mathcal{S}^n \setminus \mathcal{S}^2$ because the set function

$$A_1 \times \dots \times A_{n-1} \mapsto \text{Tr}\big(E_H(A_1)V \dots E_H(A_{n-1})V\big)$$

defined on rectangular sets of $\mathbb{R}^{n-1}$ can fail to extend to a countably additive measure of bounded variation on the Borel σ-algebra of $\mathbb{R}^{n-1}$ [74, Section 4].

Discussion of further properties of the spectral shift functions and related open questions can be found in [192, 194, 195].

Unsummable Perturbations

The condition $V \in \mathcal{S}^n$ is not satisfied by typical perturbations of differential operators, so different restrictions, including those appearing below, are imposed in the setting of differential operators.

The next result is obtained in [50, Theorem 4.6] (see also [50, Remark 4.8]).

Theorem 5.5.4 *Let $H \in \mathcal{D}_{sa}$ and $V \in \mathcal{B}_{sa}(\mathcal{H})$ be such that*

$$(H - iI)^{-1}V \in \mathcal{S}^n, \tag{5.5.13}$$

$n \in \mathbb{N}$, $n \geq 2$. *Then there exist $\eta_n = \eta_{n,H,V} \in L^1\left(\mathbb{R}, \frac{d\lambda}{1+\lambda^2}\right)$ and a constant $c_n > 0$ such that*

$$\|\eta_n\|_{L^1\left(\mathbb{R}, \frac{d\lambda}{1+\lambda^2}\right)} \leq c_n \left(1 + \|V\|\right)^{n-1} \|(H - iI)^{-1}V\|_n^n$$

and

$$\mathrm{Tr}\left(f(H + V) - f(H) - \sum_{k=1}^{n-1} \frac{1}{k!} \frac{d^k}{ds^k} f(H + sV)\Big|_{s=0} \right)$$
$$= \int_{\mathbb{R}} \frac{d^{n-1}}{dt^{n-1}}\left((t - i)^{2n} f'(t) \right) \eta_n(t)\, dt,$$

for every $f \in \mathrm{span}\left\{ \lambda \mapsto \frac{1}{(\lambda - z)^j} : \ j \in \mathbb{N},\ j \geq 2n,\ \mathrm{Im}\,(z) > 0 \right\}$.

The following result is established in [114] and [166, Theorem 3.5] in the cases $n = 1$ and $n \geq 2$, respectively.

Theorem 5.5.5 *Let $H \in \mathcal{D}_{sa}$ and $V \in \mathcal{B}_{sa}(\mathcal{H})$ be such that*

$$(H + V - iI)^{-1} - (H - iI)^{-1} \in \mathcal{S}^n, \tag{5.5.14}$$

$n \in \mathbb{N}$. *Then, there exist $\eta_n = \eta_{n,H,V} \in L^1\left(\mathbb{R}, \frac{d\lambda}{(1+\lambda^2)^{\frac{n}{2}}}\right)$ and a constant $c_n > 0$ such that*

$$\|\eta_n\|_{L^1\left(\mathbb{R}, \frac{d\lambda}{(1+\lambda^2)^{\frac{n}{2}}}\right)} \leq c_n \|(H + V - iI)^{-1} - (H - iI)^{-1}\|_n^n$$

and

$$\mathrm{Tr}\left(f(H+V)-f(H)-\sum_{k=1}^{n-1}\sum_{\substack{j_1,\dots,j_k\in\{1,\dots,n-1\}\\ j_1<\cdots<j_k}} T_{f^{[k]}}(V_{j_1}, V_{j_2-j_1},\dots, V_{j_k-j_{k-1}})\right)$$

$$=\int_{\mathbb{R}}\frac{d^{n-1}}{d\lambda^{n-1}}\left((t-i)^n f'(t)\right)\eta_n(t)\,dt \qquad (5.5.15)$$

holds for every bounded rational function f with poles in the upper half plane. Here

$$V_p:=\left((1-V(H+V-iI)^{-1})V(H-iI)^{-1}\right)^{p-1}(1-V(H+V-iI)^{-1})V$$

for $p \in \mathbb{N}$.

Remark 5.5.6

(i) Alternative trace formulas for bounded rational functions on $\mathbb{R}$ without restriction on location of their poles, but with more complicated approximating expressions are established in [191]. The proof involves trace formulas, where the approximating expressions are analogs of Taylor polynomials evaluated along multiplicative paths of unitaries, obtained in Theorem 5.5.10 below.

(ii) Another result similar to the one of Theorem 5.5.5 in the case $n = 2$ is established in [136]. Yet another condition,

$$(H+V-iI)^{-m}-(H-iI)^{-m}\in\mathcal{S}^1,$$

for odd $m \in \mathbb{N}$, is handled in [215].

We note that the intricate approximating expression

$$f(H)+\sum_{k=1}^{n-1}\sum_{\substack{j_1,\dots,j_k\in\{1,\dots,n-1\}\\ j_1<\cdots<j_k}} T_{f^{[k]}}(V_{j_1}, V_{j_2-j_1},\dots, V_{j_k-j_{k-1}}) \qquad (5.5.16)$$

in (5.5.15) cannot be replaced with the noncommutative analog of a Taylor polynomial

$$f(H)+\sum_{k=1}^{n-1}\frac{1}{k!}\frac{d^k}{ds^k}f(H+sV)\Big|_{s=0}=f(H)+\sum_{k=1}^{n-1}T_{f^{[k]}}(\underbrace{V,\dots,V}_{k\text{ times}}) \qquad (5.5.17)$$

appearing in (5.5.2) because the trace of $f(H+V)-(5.5.17)$ is generally undefined under the assumption (5.5.14).

Remark 5.5.7

(i) The conditions considered in Theorems 5.5.1, 5.5.4, and 5.5.5 are related by

$$(5.5.1) \Rightarrow (5.5.13) \Rightarrow (5.5.14),$$

which follows immediately from resolvent identities.

(ii) The conditions (5.5.13) and (5.5.14) naturally arise in the study of Dirac as well as Schrödinger operators and their perturbations, as demonstrated in [50, 166]. Examples involving fractional powers of Laplacians are considered in [162].

(iii) The proofs of Theorems 5.5.4 and 5.5.5 rely on tricks that reduce the case of unsummable perturbations to the case of summable perturbations. More specifically, the proof of Theorem 5.5.5 uses a change of operator variables (see Theorem 4.3.21) in a multiple operator integral that transfers the remainder $f(H + V) - (5.5.16)$ for H and $H + V$ satisfying (5.5.14) to the nth Taylor remainder for pairs of contractions, which are the Cayley transforms of the operators H and $H + V$, respectively, and whose difference is an element of $\mathcal{S}^n$. The respective result for contractions is stated in Theorem 5.5.12 below. The proof of Theorem 5.5.4 benefits from a method of a summable weight applied to the difference of the approximating expressions (5.5.16)–(5.5.17).

Below we consider another case of unsummable perturbations, where a summability restriction is imposed on the resolvent of H, but V is assumed to be an arbitrary bounded operator. The following result is established in [23, Theorem 2.5], [189, Theorem 3.10], and [193] for $n = 1$, $n = 2$, and $n \geqslant 3$, respectively. Its proof follows from the estimate for the remainder stated in Theorem 5.4.12.

Theorem 5.5.8 *Let $H \in \mathcal{D}_{sa}$ have compact resolvent, $V \in \mathcal{B}_{sa}(\mathcal{H})$, $n \in \mathbb{N}$. Then, there exists a unique real-valued locally integrable function $\eta_{n,H,V}$ such that*

$$\mathrm{Tr}(f(H + V)) = \sum_{k=0}^{n-1} \frac{1}{k!} \mathrm{Tr}\left(\frac{d^k}{ds^k} f(H + sV) \Big|_{s=0} \right) + \int_{\mathbb{R}} f^{(n)}(t)\, \eta_{n,H,V}(t)\, dt,$$

(5.5.18)

for $f \in C_c^{n+1}(\mathbb{R})$. If $n \geqslant 2$, then the function $\eta_{n,H,V}$ satisfies the bound

$$\int_{[a,b]} |\eta_{n,H,V}(t)|\, dt$$

$$\leqslant \|V\|^n \left(2^n(n+1) + c_n\right) (b - a + 1)^n \left(1 + \Omega_{a,b,H,V}\right)$$

$$\times \left(\Omega_{a_\epsilon, b_\epsilon, H, V} + \sqrt{2}\, (b_\epsilon - a_\epsilon + 1)^{3/2} \max_{1 \leqslant k \leqslant n} \frac{\|\varphi_\epsilon^{(k+1)}\|_\infty}{k!} \right),$$

for every $\epsilon > 0$. Here $c_n > 0$, $\Omega_{a,b,H,V}$ is such that

$$\sup_{t \in [0,1]} \mathrm{Tr}\big(E_{H+tV}((a,b))\big) \leqslant \Omega_{a,b,H,V} \qquad (5.5.19)$$

and

$$\mu_{\Omega_{a,b,H,V}}\big((I + H^2)^{-1}\big) \leqslant \frac{1}{(1 + \max\{a^2, b^2\})(1 + \|V\| + \|V\|^2)}$$

hold and φ_ϵ is a smoothening of the indicator function of (a,b) satisfying $\sqrt[4]{\varphi_\epsilon} \in C_c^\infty((a_\epsilon, b_\epsilon))$, $\varphi_\epsilon|_{(a,b)} \equiv 1$, and $0 \leqslant \varphi_\epsilon \leqslant 1$, where $a_\epsilon = a - \epsilon$, $b_\epsilon = b + \epsilon$.

Remark 5.5.9

(i) Examples of operators with compact resolvents include differential operators on compact Riemannian manifolds (see, e.g., [29, Chapter 3, Section B], [100, Chapter 3, Section 6]).

(ii) The proof of Theorem 5.5.8 employs an auxiliary function φ_ϵ to create summable perturbations that are amiable to methods of the previous subsection.

5.5.2   *Spectral Shift Function for Nonself-adjoint Operators*

The question of validity of (5.5.2) for nonself-adjoint operators H and $H + V$ is also of interest in perturbation theory. In particular, trace formulas for unitaries and contractions have been involved in derivation of trace formulas for unbounded self-adjoint and dissipative operators.

The following trace formulas are based on analogs of Taylor polynomials evaluated along multiplicative paths of unitaries; they are established in [114, Theorem 2] and [191, Theorem 4.4] for $n = 1$, $n \geqslant 2$, respectively.

Theorem 5.5.10 *Let $n \in \mathbb{N}$, U be a unitary, $A = A^* \in \mathcal{S}^n$, $U(t) = e^{itA}U$. Then, there exist a function $\eta_n = \eta_{n,A,U} \in L^1[0, 2\pi]$ and $c_n > 0$ such that*

$$\|\eta_n\|_1 \leqslant c_n \|A\|_n^n$$

and

$$\mathrm{Tr}\left(\varphi(e^{iA}U) - \varphi(U) - \sum_{k=1}^{n-1} \frac{1}{k!} \frac{d^k}{ds^k} \varphi(U(s))\big|_{s=0}\right) = \int_0^{2\pi} \varphi^{(n)}(e^{it})\eta_n(t)\,dt,$$

$$(5.5.20)$$

for every $\varphi \in C^n(\mathbb{T})$ such that $\varphi(z) = \sum_{j=-\infty}^{\infty} a_j z^j$, where $\sum_{j=-\infty}^{\infty} |a_j||j|^n < \infty$.

Remark 5.5.11

(i) The respective η_1 is real-valued and unique up to an additive constant. The maximal class of functions satisfying (5.5.20) with $n = 1$ is the class of operator Lipschitz functions on $\mathbb{T}$, as established in [10].

(ii) Another version of (5.5.20) with $n = 2$ is obtained in [136] and extended to φ in the Besov class of functions $B_{\infty 1}^2(\mathbb{T})$ in [145].

(iii) The result of Theorem 5.5.10 with $n \geqslant 2$ for functions φ analytic on the unit disc and such that $\varphi^{(n)}$ on $\mathbb{T}$ is given by an absolutely convergent Taylor series is obtained [168, Theorem 4.1].

In [118], an analog of the trace formula (5.5.20) with $n = 1$ was obtained for bounded operators, with the integral over $[0, 2\pi]$ replaced by an integral over a planar contour containing $\sigma(H) \cup \sigma(H + V)$ in its interior. Later in [129], this contour was changed to the unit circle in the case when H and $H + V$ are contractions. Analogous higher order trace formulas for Taylor remainders evaluated along linear paths of contractions were established in [160, Theorem 1] and [164, Theorem 1.3] for $n = 2$ and $n \geqslant 3$, respectively. These results are summarized in the theorem below.

Theorem 5.5.12 *Let $n \in \mathbb{N}$, H and $H + V$ be contractions, and assume that $V \in S^n$. Then, there exist a function $\eta_n = \eta_{n,H,V}$ in $L^1(\mathbb{T})$ and $c_n > 0$ such that*

$$\|\eta_n\|_1 \leqslant c_n \|V\|_n^n$$

and

$$\mathrm{Tr}\left(f(H + V) - \sum_{k=0}^{n-1} \frac{1}{k!} \frac{d^k}{ds^k} f(H + sV)\big|_{s=0}\right) = \int_{\mathbb{T}} f^{(n)}(z)\,\eta_n(z)\,dz,$$

(5.5.21)

for every f analytic on a disc centered at 0 of radius greater than 1.

Remark 5.5.13

(i) The spectral shift function η_n satisfying Theorem 5.5.12 is determined uniquely only up to an analytic term, that is, the equivalence class of η_n in the quotient space $L^1(\mathbb{T})/H^1(\mathbb{T})$ is uniquely determined. The proof in the case $n \geqslant 2$ relies on the estimate (5.3.13) and is similar to the proof of Theorem 5.5.1, but it requires more careful integration by parts and approximation arguments.

(ii) The result of Theorem 5.5.12 with $n = 1$ is extended in [129] to operator Lipschitz functions f that are in $\mathcal{A}(\mathbb{D})$. Earlier attempts to extend (5.5.21) with $n = 1$ to more general functions f resulted in consideration of selected pairs of contractions and brought to modification of (5.5.21) with passage to a more general type of integration [2, 176–179]. For the relevant discussion,

we refer the reader to [127, 188]; see also [130] for the proof of the absolute
continuity of the spectral shift based on dilation theory.
(iii) The result of Theorem 5.5.12 with $n \geqslant 2$ is written in [160, 164] for polynomials f, but it readily extends to the case of analytic functions.

Some physical systems can be modeled by dissipative or accumulative operators.
We recall that an operator is called maximal dissipative (respectively, accumulative)
if it is closed, densely defined and its quadratic form has a nonnegative (respectively,
nonpositive) imaginary part. We refer the reader to [3, 12, 116, 127, 128, 130, 134,
135, 175, 180] for results on trace formulas for dissipative operators and to [127,
188] for the history of the subject.

Problem 5.5.14 Does there exist an analog of the spectral shift function for a pair
of normal operators whose difference belongs to $\mathcal{S}^1$?

A partial result in this direction is obtained in [190, Theorem 3.8].

Theorem 5.5.15 *Let $A_j = A_j^*$ and $B_j = B_j^*$ be contractions such that $B_j - A_j \in$
$\mathcal{S}^1$, $j = 1, 2$. Assume that $[A_1, A_2] = 0$, $[B_1, B_2] = 0$, and $[B_1 - A_1, B_2 - A_2] = 0$.
Then, there exist real-valued measures μ_1, μ_2 on $[-1, 1] \times [-1, 1]$ such that*

$$\|\mu_j\| \leqslant \|B_j - A_j\|_1,$$

$$\mu_j([-1, 1] \times [-1, 1]) = \mathrm{Tr}(B_j - A_j), \quad j = 1, 2,$$

and

$$\mathrm{Tr}\big(f(B_1, B_2) - f(A_1, A_2)\big)$$

$$= \int_{[-1,1] \times [-1,1]} \frac{\partial f}{\partial z_1}(z_1, z_2) \, d\mu_1(z_1, z_2)$$

$$+ \int_{[-1,1] \times [-1,1]} \frac{\partial f}{\partial z_2}(z_1, z_2) \, d\mu_2(z_1, z_2),$$

for every polynomial f of two variables. Moreover, the measures μ_1 and μ_2 satisfy

$$\int_{[-1,1] \times [-1,1]} g(z_j) \, d\mu_j(z_1, z_2) = \int_{-1}^{1} g(t) \, \eta_{1, A_j, B_j - A_j}(t) \, dt,$$

*for every polynomial g of one variable, where $\eta_{1, A_j, B_j - A_j}$ is the spectral shift
function for the pair (A_j, B_j) provided by Theorem 5.5.1, $j = 1, 2$.*

It would be interesting to establish absolute continuity of the measures μ_1 and
μ_2 provided by Theorem 5.5.15 and remove the extra commutativity restriction on
the operators $[B_1 - A_1, B_2 - A_2] = 0$.

5.5.3 Spectral Shift Measure in the Setting of von Neumann Algebras

The result of Theorem 5.5.1 holds for $H\eta M_{sa}$ and $V = V^* \in \mathcal{L}^n(\mathcal{M}, \tau)$, where $\mathcal{M}$ is a semifinite von Neumann algebra equipped with a normal faithful semifinite trace τ. The case $n = 1$ was established first for a bounded operator H in [47] and then for an unbounded operator in [22]. The case $n = 2$ is due to [74, 186] and $n \geqslant 3$ is due to [163]. Properties of the first and second order spectral shift functions in semi-finite von Neumann algebras are discussed in [184, 185]. The result of Theorem 5.5.8 holds under the generalized assumption that the resolvent of H is τ-compact [23, 193]; in this case, Tr in (5.5.18) and (5.5.19) is replaced with τ. The strategy of the proofs is as described in Sect. 5.5.1; this strategy can be implemented because noncommutative L^p-spaces have much in common with Schatten ideals (see, e.g., [154]).

The spectral shift measure has also been studied in the setting of general symmetrically normed ideals of semifinite von Neumann algebras. We state major results below.

Hypotheses 5.5.16 *Let* $n \in \{1, 2\}$, $\mathcal{M}$ *be a semifinite von Neumann factor,* $\mathcal{I}$ *a symmetrically normed ideal of* $\mathcal{M}$ *with the ideal norm* $\|\cdot\|_\mathcal{I}$ *and a trace* $\tau_\mathcal{I}$ *bounded with respect to* $\|\cdot\|_\mathcal{I}$. *Consider a set* Ω, *operators* $H\eta M$, $V \in \mathcal{I}^{1/n}$, *and a space* $\mathcal{F}$ *of functions that satisfy one of the following assertions:*

(i) $\Omega = \mathrm{conv}(\sigma(H) \cup \sigma(H + V))$, $H, V \in M_{sa}$, $\mathcal{F} = C^{n+2}(\mathbb{R})$;
(ii) $\Omega = \mathbb{R}$, H *and* $H + V$ *are maximal dissipative operators, and* $\mathcal{F} = \mathcal{R}_-$;
(iii) $\Omega = \mathbb{T}$, $\|H\| \leqslant 1$, $\|H + V\| \leqslant 1$, *and* $\mathcal{F}$ *is the set of all functions that are analytic on a disc centered at 0 and of radius strictly larger than 1.*

The following two results are established in [75, Theorems 3.4 and 5.3].

Theorem 5.5.17 *Let* Ω, H, V *and* $\mathcal{F}$ *satisfy Hypotheses 5.5.16 with* $n = 1$. *Then, there exists a (countably additive, complex) measure* $\nu_1 = \nu_{1,H,V}$ *on* Ω *such that*

$$\|\nu_1\| \leqslant \min\left\{\tau_\mathcal{I}(|\mathrm{Re}\,(V)|) + \tau_\mathcal{I}(|\mathrm{Im}\,(V)|), \|\tau_\mathcal{I}\|_{\mathcal{I}^*} \cdot \|V\|_\mathcal{I}\right\}$$

and

$$\tau_\mathcal{I}\big(f(H + V) - f(H)\big) = \int_\Omega f'(\lambda)\,\nu_1(d\lambda),$$

for all $f \in \mathcal{F}$. *If Hypotheses 5.5.16(i) are satisfied, then the measure* ν_1 *is real and unique.*

Remark 5.5.18 When $(\mathcal{I}, \tau_\mathcal{I})$ is the dual Macaev ideal with the Dixmier trace, the measure ν_1 can be of any type [75, Theorem 4.4], as distinct from the case of an absolutely continuous measure in Theorem 5.5.1. Moreover, we do not have an explicit formula for ν_1 in the case of a general trace $\tau_\mathcal{I}$. Derivation of an explicit formula for ν_1 in the case $\mathcal{I} = \mathcal{S}^1$, $H = H^*$, and $V = V^*$ relies on the fact that $\mathrm{Tr}(E_H(\cdot)V)$ is a (countably-additive) measure, while the set function $\mathrm{Tr}_\omega(E_H(\cdot)V)$ can fail to be countably-additive (see [75, Section 3]).

Theorem 5.5.19 *Let* Ω, H, V *and* $\mathcal{F}$ *satisfy Hypotheses 5.5.16 with* $n = 2$. *Then, there exists a (countably additive, complex) measure* $\nu_2 = \nu_{2,H,V}$ *on* Ω *such that*

$$\|\nu_2\| \leqslant \frac{1}{2}\,\tau_I(|V|^2)$$

and

$$\tau_I\left(f(H+V) - f(H) - \frac{d}{ds}f(H+sV)\big|_{s=0}\right) = \int_\Omega f''(\lambda)\,\nu_2(d\lambda),$$

for every $f \in \mathcal{F}$.

The following higher order result for a self-adjoint operator V in the root of the dual Macaev ideal is established in [167, Theorems 1.2 and 1.3].

Theorem 5.5.20 *Let* $n \in \mathbb{N}$, $H\eta\mathcal{M}_{sa}$, $V = V^* \in (\mathcal{L}^{(1,\infty)})^{1/n}$, *and let* $\tau_{\mathcal{L}^{(1,\infty)}}$ *be a trace on* $\mathcal{L}^{(1,\infty)}$ *continuous with respect to* $\|\cdot\|_{\mathcal{L}^{(1,\infty)}}$. *Then, there exists a (countably additive, complex) measure* $\nu_n = \nu_{n,H,V}$ *and* $c_n > 0$ *such that*

$$\|\nu_n\| \leqslant c_n\,\tau_{\mathcal{L}^{(1,\infty)}}(|V|^n)$$

and

$$\tau_{\mathcal{L}^{(1,\infty)}}\left(f(H+V) - \sum_{k=0}^{n-1}\frac{1}{k!}\frac{d^k}{ds^k}f(H+sV)\big|_{s=0}\right) = \int_{\mathbb{R}} f^{(n)}(t)\,\nu_n(dt),$$

$$(5.5.22)$$

for every $f \in \mathcal{R}$. *Moreover, if* H *is bounded, then the formula (5.5.22) holds for every Schwartz function* f.

Remark 5.5.21 Major components in the proofs of Theorems 5.5.17, 5.5.19, and 5.5.20 are analogs of the estimates similar to (5.3.13). However, presence of a singular component in the trace τ_I requires more careful treatment of the operator derivatives than in the case of the normal trace Tr.

5.6   Spectral Flow

Spectral flow stems from the work of M. F. Atiyah, V. K. Patodi, I. M. Singer [20], where it is introduced primarily in a topological sense. In 1974 it was suggested by I. M. Singer that the spectral flow can be expressed as an integral of one-form, and this idea was pursued in [45, 46, 84]. The spectral flow in the context of von Neumann algebras was considered by J. Phillips in [150, 151], where an analytic approach was taken. This analytic approach is logically equivalent to the one of [20], but technically simpler to work with. The relevant details and definitions can be found in [25]. The spectral flow has connections with noncommutative geometry (see [25] and references cited therein) as well as perturbation theory (see Theorem 5.6.4).

For the rest of this subsection we assume that $\mathcal{M}$ is a semifinite von Neumann algebra equipped with a normal faithful semifinite trace τ. Let $D_0 \eta \mathcal{M}_{sa}$ have τ-compact resolvent, $V \in \mathcal{M}_{sa}$, and denote

$$D_r = D_0 + rV.$$

Let $\mathrm{sf}(D_0, D_1) = \mathrm{sf}(\{D_r\}_r)$ denote the spectral flow from D_0 to D_1, which is independent of a continuous path joining the points D_0 and D_1 by the homotopy invariance property of the spectral flow. By definition [25, Section 7.2], the spectral flow for a pair of unbounded τ-Fredholm operators is

$$\mathrm{sf}(D_0, D_1) := \mathrm{sf}(\{F_r\}_r),$$

where $r \to F_r$ is a norm-continuous path of bounded τ-Fredholm operators

$$F_r := D_r (I + D_r^2)^{-1/2}.$$

In its turn,

$$\mathrm{sf}(\{F_r\}_r) = \sum_{i=1}^{k} ec\left(\chi_{[0,\infty)}(F_{t_{i-1}}), \chi_{[0,\infty)}(F_{t_i})\right)$$

for a suitable partition $0 = t_0 < t_1 < \cdots < t_{k-1} < t_k = 1$, where $ec(P, Q)$ denotes the essential codimension of the projection P in Q (see, e.g., [25, Definitions 4.1 and 4.2]).

The results stated below are due to [23]; they generalize the respective results of [45, 46] and rely on methods of double operator integrals in addition to traditional analytic methods for the spectral flow.

The next result is obtained in [23, Theorems 3.23].

Theorem 5.6.1 *Let $D_0 \eta \mathcal{M}_{sa}$ have τ-compact resolvent, let $f \in L^1(\mathbb{R})$ be a nonnegative function such that $f(D_r) \in \mathcal{L}^1(\mathcal{M}, \tau)$ for all $r \in [0, 1]$ and $r \mapsto \|f(D_r)\|_1$ is integrable on $[0, 1]$. If D_0 and D_1 are unitarily equivalent, then*

$$\mathrm{sf}(D_0, D_1) = \left(\int_{-\infty}^{\infty} f(t)\, dt\right)^{-1} \int_0^1 \tau(V f(D_r))\, dr.$$

By [23, Proposition 1.20], every function $0 \leqslant f \in C_c^2(\mathbb{R})$ satisfies the assumptions of Theorem 5.6.1. The spectral flow in Theorem 5.6.1 is calculated as an integral of 1-form, the property established in [23, Propositions 3.3, 3.5, 3.8] and summarized below.

Theorem 5.6.2 *Let $D_0 \eta M_{sa}$ have τ-compact resolvent, $V \in M_{sa}$, $D \in D_0 + M_{sa}$, and $f \in C_c^3(\mathbb{R})$. Then,*

$$\alpha_D^f(V) := \tau(Vf(D))$$

defines a closed exact 1-form α_D^f and

$$\alpha_D^f(X) = d\theta_D^f(X), \quad X \in M, \tag{5.6.1}$$

where

$$\theta_D^f(V) := \int_0^1 \tau(Vf(D_r))\, dr.$$

Proof The properties of α_D^f and the formula (5.6.1) are derived by double operator integral techniques. We demonstrate involvement of these techniques in the calculation of $d\theta_D^f(X)$ for $X = X^*$.

By definition of the derivative,

$$d\theta_D^f(X) = \frac{d}{ds}\theta_{D+sX}^f\big|_{s=0}$$

$$= \lim_{s\to 0} \frac{1}{s} \int_0^1 \tau\big((V+sX)f(D_r + srX) - Vf(D_r)\big)\, dr$$

$$= \lim_{s\to 0} \int_0^1 \tau\big(Xf(D_r + srX)\big)\, dr$$

$$+ \lim_{s\to 0} \frac{1}{s} \int_0^1 \tau\big(V(f(D_r + srX) - f(D_r))\big)\, dr.$$

Hence, by continuity of $s \mapsto \tau\big(Xf(D_r + srX)\big)$,

$$d\theta_D^f(X) = \int_0^1 \tau\big(Xf(D_r)\big)\, dr + \lim_{s\to 0} \frac{1}{s} \int_0^1 \tau\big(V(f(D_r + srX) - f(D_r))\big)\, dr. \tag{5.6.2}$$

To finish the proof, we apply technical modifications of the double operator integral methods adjusted to the setting of operators with τ-compact resolvents. More precisely, by the perturbation formula [23, Proposition 1.18(i)], linearity and continuity of the double operator integrals on noncommutative L^p-spaces [23, Lemma 1.22], and the cyclicity of the trace and [23, Lemma 3.2],

$$\lim_{s \to 0} \frac{1}{s} \int_0^1 \tau\big(V(f(D_r + srX) - f(D_r))\big)\, dr$$

$$= \lim_{s \to 0} \int_0^1 \tau\big(V T_{f^{[1]}}^{D_r + srX, D_r}(rX)\big)\, dr$$

$$= \int_0^1 \tau\big(V T_{f^{[1]}}^{D_r, D_r}(rX)\big)\, dr = \int_0^1 \tau\big(X T_{f^{[1]}}^{D_r, D_r}(V)\big)\, r\, dr. \qquad (5.6.3)$$

From (5.6.2) and (5.6.3) we obtain

$$d\theta_D^f(X) = \tau\Big(X \int_0^1 \big(f(D_r) + r\, T_{f^{[1]}}^{D_r, D_r}(V)\big)\, dr\Big). \qquad (5.6.4)$$

By continuity of $r \mapsto f(D_r) + r T_{f^{[1]}}^{D_r, D_r}(V)$ in $\mathcal{L}^1(\mathcal{M}, \tau)$, perturbation formula [23, Proposition 1.18(i)], and the integral representation for the increment of the operator function

$$f(D_{r_1}) - f(D_{r_0}) = \int_{r_0}^{r_1} T_{f^{[1]}}^{D_r, D_r}(V)\, dr$$

[23, (12)], one calculates

$$\int_0^1 \big(f(D_r) + r T_{f^{[1]}}^{D_r, D_r}(V)\big)\, dr = \mathcal{L}^1\text{-}\lim_{n \to \infty} \frac{1}{n} \sum_{j=1}^n \Big(f(D_{r_{j-1}}) + \frac{j}{n} T_{f^{[1]}}^{D_{r_j}, D_{r_j}}(V)\Big)$$

$$= f(D_1), \qquad (5.6.5)$$

where $0 = r_0 < r_1 < \cdots < r_n = 1$ is a partition of $[0, 1]$ into segments of length $1/n$. Combining (5.6.4) and (5.6.5) completes the proof. $\qquad \square$

When the end points of a path are not unitarily equivalent, the spectral flow is calculated in [23, Theorem 3.28] as follows.

Theorem 5.6.3 *Let $D_0 \eta \mathcal{M}_{sa}$ satisfy $\tau\big(\exp(-tD_0^2)\big) < \infty$ for all $t > 0$. Then,*

$$\mathrm{sf}(D_0, D_1) = \sqrt{\frac{\varepsilon}{\pi}} \int_0^1 \tau(V \exp(-\varepsilon D_r^2))\, dr$$

$$+ \frac{1}{2}(\eta_\varepsilon(D_1) - \eta_\varepsilon(D_0)) + \frac{1}{2}\big(\tau\big(P_{\ker(D_1)}\big) - \tau\big(P_{\ker(D_0)}\big)\big),$$

where the truncated eta-invariant η_ε is defined by

$$\eta_\varepsilon(D) = \frac{1}{\sqrt{\pi}} \int_\varepsilon^\infty \tau(D \exp(-tD^2))t^{-1/2}\, dt.$$

The spectral flow is closely related to the spectral shift function, as established in the next result due to [23, Theorems 3.18]. Let sf(λ, D_0, D_1) denote the spectral flow from $D_0 - \lambda I$ to $D_1 - \lambda I$.

Theorem 5.6.4 *Let F_0, $F_1\eta M_{sa}$ be τ-Fredholm operators with τ-compact difference $F_1 - F_0$ and such that $F_i^2 - I$ is τ-compact, $i = 1, 2$. Then,*

$$\mathrm{sf}(\lambda, F_0, F_1) = \xi_{F_0, F_1 - F_0}(\lambda) + \frac{1}{2}\left(\tau\left(P_{\ker(F_1 - \lambda I)}\right) - \tau\left(P_{\ker(F_0 - \lambda I)}\right)\right),$$

for all $\lambda \in (-1, 1)$, where $\xi_{F_0, F_1 - F_0}$ is the spectral shift function discussed in Sect. 5.5.3.

The result of Theorem 5.6.4 is based on the following representation of the spectral shift function due to [23, Lemma 2.8], which extends the result of [37].

Lemma 5.6.5 *Let $D_0\eta M_{sa}$ and $V \in M_{sa}$. If $f : \mathbb{R} \to \mathbb{C}$ is a bounded compactly supported function, then*

$$\int_{\mathbb{R}} f(\lambda)\xi_{D_0, V}(\lambda)\, d\lambda = \int_0^1 \tau(Vf(D_r))\, dr.$$

5.7 Quantum Differentiability

A quantized differential was introduced by A. Connes to replace a differential calculus in noncommutative differential geometry by an operator theoretic notion involving a commutator. The details on this account can be found in [57, Chapter 4]. The asymptotic behavior of the singular values of a quantized derivative determines the dimension of the infinitesimal in the quantized calculus. In this section we discuss a characterization of the set of functions for which the quantized derivative belongs to the weak space $\ell^{d,\infty}$, which is analogous to behaviour of differential forms.

Let $f \in L^\infty(\mathbb{R}^d)$ and M_f be the operator of pointwise multiplication by f on the space $L^2(\mathbb{R}^d)$, $d \in \mathbb{N}$. If $d > 1$, let $N = 2^{\lfloor d/2 \rfloor}$ and let $\{\gamma_j\}_{j=1}^d$ be $N \times N$ gamma matrices, which satisfy $\gamma_j^2 = 1$, $\gamma_j^* = \gamma_j$, $\gamma_j\gamma_k = -\gamma_k\gamma_j$, $j \neq k$. Let

$$D = \sum_{j=1}^d \gamma_j \otimes \left(-i\frac{\partial}{\partial x_j}\right)$$

be an operator densely defined in $\mathbb{C}^N \otimes L^2(\mathbb{R}^d)$. Let sgn (D) be defined by the functional calculus. The *quantized derivative* of f is defined as an operator in $\mathcal{B}(\mathbb{C}^N \otimes L^2(\mathbb{R}^d))$ by the formula

$$\text{d}f := i[\text{sgn}(D), I \otimes M_f].$$

The functions $f \in L^\infty(\mathbb{R})$ for which $\text{d}f \in \mathcal{S}^{1,\infty}$ are characterized in [144, Chapter 6, Theorem 4.4]. The functions $f \in L^\infty(\mathbb{R}^d)$ for which $\text{d}f \in \mathcal{S}^{d,\infty}$ are characterized in [172, Corollary 28, Theorem 3.4] in terms of the mean oscillation of f. We also note that $\text{d}f \in \mathcal{S}^d$ if and only if f is a constant [92].

The following result was stated in [58, p. 679], and a different complete proof was provided in [124, Theorem 1].

Theorem 5.7.1 *Let $d \in \mathbb{N}$, $d > 1$, and $f \in L^\infty(\mathbb{R}^d)$. Then, $\text{d}f \in \mathcal{S}^{d,\infty}(\mathbb{C}^N \otimes L^2(\mathbb{R}^d))$ if and only if $\nabla f \in L^d(\mathbb{R}^d, \mathbb{C}^d)$. Moreover, there exist constants $c_d, C_d > 0$ such that*

$$c_d \|\nabla f\|_{L^d(\mathbb{R}^d, \mathbb{C}^d)} \leqslant \|\text{d}f\|_{d,\infty} \leqslant C_d \|\nabla f\|_{L^d(\mathbb{R}^d, \mathbb{C}^d)}.$$

Proof We outline major ideas of the proofs.

The necessary part of the theorem is derived from the trace formula

$$\varphi(|\text{d}f|^d) = a_d \int_{\mathbb{R}^d} \|\nabla f(x)\|_2^d \, dx,$$

where φ is a continuous normalized trace on $\mathcal{S}^{1,\infty}$ and $a_d > 0$, which is established in [124, Theorem 17].

The sufficient condition is confirmed with involvement of double operator integration. If f is a Schwartz function, then, by Theorem 5.1.4 applied to the function

$$g(t) = t(1 + t^2)^{-\frac{1}{2}},$$

we have

$$[D(I + D^2)^{-\frac{1}{2}}, I \otimes M_f] = T^{D,D}_{g^{[1]}}([D, I \otimes M_f]).$$

Based on properties of double operator integrals discussed in Sect. 3.3.5, it is proved in [124, Lemma 10] that $T^{D,D}_{g^{[1]}} \in \mathcal{B}(\mathcal{S}^1)$ and $T^{D,D}_{g^{[1]}} \in \mathcal{B}(\mathcal{B}(\mathbb{C}^N \otimes L^2(\mathbb{R}^d)))$. By interpolation [69], $T^{D,D}_{g^{[1]}} \in \mathcal{B}(\mathcal{S}^{d,\infty})$ and there exists $b_d > 0$ such that

$$\left\|[D(I + D^2)^{-\frac{1}{2}}, I \otimes M_f]\right\|_{d,\infty} \leqslant b_d \|\nabla f\|_{L^d(\mathbb{R}^d, \mathbb{C}^d)}. \tag{5.7.1}$$

It is proved in [124, Lemma 7] that

$$\left\|[\operatorname{sgn}(D) - D(I + D^2)^{-\frac{1}{2}}, I \otimes M_f]\right\|_{d,\infty} \leqslant C_{p,d} \begin{cases} \|f\|_{\ell^p(L^2)} & \text{if } 1 \leqslant p < 2 \\ \|f\|_p & \text{if } p \geqslant 2. \end{cases}$$

(5.7.2)

Applying (5.7.1), (5.7.2), a dilation argument, and approximations completes the proof.   □

5.8 Differentiation of Noncommutative L^p-Norms

Let $\mathcal{M} \subseteq \mathcal{B}(\mathcal{H})$ be an arbitrary von Neumann algebra and $L^p_{Haag}(\mathcal{M})$, $1 \leqslant p < \infty$, the Haagerup L^p-space associated with $\mathcal{M}$ (see the definition in Sect. 2.7). In this section we discuss resolution of the question on differentiability of the norm $\| \cdot \|_{L^p_{Haag}}$ on $L^p_{Haag}(\mathcal{M})$ that was suggested by G. Pisier and Q. Xu in [154], generalizing the earlier question of N. Tomczak-Jaegermann [204].

The following result is proved in [170, Theorem 1], extending the analogous result of [161, Theorem 15] for the classical noncommutative L^p-space.

Theorem 5.8.1 *The function* $\| \cdot \|^p_{L^p_{Haag}}$ *is*

(i) infinitely many times Taylor-Fréchet differentiable whenever p is an even integer;

(ii) $(p-1)$-times Taylor-Fréchet differentiable whenever p is an odd integer;

(iii) $\lfloor p \rfloor$-times Taylor-Fréchet differentiable whenever p is not an integer.

Remark 5.8.2 The result of Theorem 5.8.1 is sharp. Indeed, if $\mathcal{M}$ is a von Neumann algebra of type I, then $L^p_{Haag}(\mathcal{M})$ contains an isometric copy of the space ℓ^p. If $\mathcal{M}$ is not an algebra of type I, then $L^p_{Haag}(\mathcal{M})$ contains an isometric copy of $L^p_{Haag}(\mathcal{R})$ for the hyperfinite II_1 factor $\mathcal{R}$ (see [154, Theorem 3.5]). It is clear that $L^p_{Haag}(\mathcal{R})$ contains an isometric copy of $L^p(0, 1)$. The sharpness of Theorem 5.8.1 follows now from the differentiability properties of the norms of the classical spaces $L^p(0, 1)$ and ℓ^p (see [44, 201]).

We prove Theorem 5.8.1 in two steps, firstly the simple case of an even p and then the general case.

Proposition 5.8.3 *Let* $p \in \mathbb{N}$ *be even and* A, X *be self-adjoint elements in* $L^p_{Haag}(\mathcal{M})$. *Then,*

$$\|A + X\|^p_{L^p_{Haag}} - \|A\|^p_{L^p_{Haag}} - \sum_{k=1}^{p} \frac{1}{k!} \Delta^A_{k,p} \underbrace{\left(X, \ldots, X \right)}_{k\text{-times}} = 0,$$

(5.8.1)

where $\Delta_{k,p}^A$ are symmetric multilinear bounded functionals on $L_{Haag}^p(\mathcal{M})^{\times k}$ defined by

$$\Delta_{k,p}^A\left(X_1, \ldots, X_k\right) := \sum_{\pi \in \mathrm{Sym}_k} \mathrm{tr}\left(\sum_{\substack{s_0 + \cdots + s_k = p-k, \\ s_0, \ldots, s_k \geqslant 0}} A^{s_0} X_{\pi(1)} A^{s_1} \cdots X_{\pi(k)} A^{s_k} \right),$$

where Sym_k is the group of all permutations of the set $\{1, \ldots, k\}$ and $X_j \in L_{Haag}^p(\mathcal{M})$, $j = 1, \ldots, k$, $k = 1, \ldots, p$. Moreover,

$$\left| \Delta_{k,p}^A\left(\underbrace{X, \ldots, X}_{k\text{-times}} \right) \right| \leqslant \frac{p!}{(p-k)!} \|A\|_{L_{Haag}^p}^{p-k} \|X\|_{L_{Haag}^p}^k, \tag{5.8.2}$$

$j = 1, \ldots, k$, $k = 1, \ldots, p$.

Proof The equality (5.8.1) follows from the binomial expansion of $|A + X|^p$ and definition (2.7.3). The estimate (5.8.2) is a straightforward consequence of the Hölder inequality (2.7.5). $\qquad\square$

The first order differentiability is essentially established in [112].

Lemma 5.8.4 *For every $1 < p < \infty$, the function $\|\cdot\|_{L_{Haag}^p}^p$ is Fréchet differentiable at $A \in L_{Haag}^p(\mathcal{M})$ and the respective Fréchet derivative is given by*

$$D(\|A\|_{L_{Haag}^p}^p)(X) = p \cdot \mathrm{tr}\,(X \cdot |A|^{p-1}\mathrm{sgn}(A)),$$

where tr is the trace on $L_{Haag}^1(\mathcal{M})$ defined by (2.7.2).

Proof For $A = 0$ the assertion is trivial. Let $0 \neq A \in L_{Haag}^p(\mathcal{M})$. By [112, Lemma 3.1], the norm $\|\cdot\|_{L_{Haag}^p} : L_{Haag}^p(\mathcal{M}) \to \mathbb{R}$ is Fréchet differentiable and its Fréchet derivative is given by

$$D(\|A\|_{L_{Haag}^p})(X) = \|A\|_{L_{Haag}^p}^{1-p} \mathrm{tr}(X \cdot |A|^{p-1}\mathrm{sgn}(A)), \quad X \in L_{Haag}^p(\mathcal{M}).$$

By the above formula and the chain rule,

$$D(\|A\|_{L_{Haag}^p}^p)(X) = p \cdot \|A\|_{L_{Haag}^p}^{p-1} \|A\|_{L_{Haag}^p}^{1-p} \mathrm{tr}(X \cdot |A|^{p-1}\mathrm{sgn}(A)),$$

concluding the proof. $\qquad\square$

Below we prove Theorem 5.8.1(ii) and (iii) for a self-adjoint A.

Theorem 5.8.5 *Let* $m \geqslant 2$ *and* $p \in (m, m+1]$. *For self-adjoint* $A, X_1, \ldots, X_m \in$ $L^{p,\infty}(\mathcal{N}, \tau)$, *define*

$$\delta^A_{k,p,\phi}(X_1, \ldots, X_k) \tag{5.8.3}$$

$$:= \begin{cases} \phi(X_1 \cdot (f^p)'(A)), & k = 1 \\ \frac{1}{k!} \sum_{\pi \in \mathrm{Sym}_k} \phi\big(X_{\pi(1)} \cdot T^{A,\ldots,A}_{((f^p)')^{[k-1]}}(X_{\pi(2)}, \ldots, X_{\pi(k)})\big), & k = 2, \ldots, m, \end{cases}$$

where $\mathcal{N} = \mathcal{M} \rtimes_{\sigma^{\phi^0}} \mathbb{R}$ *is equipped with the canonical semifinite trace* τ, ϕ *is a normalized positive trace on* $L^{1,\infty}(\mathcal{N}, \tau)$ *given by (2.9.1), and* $f \in C_c(\mathbb{R})$ *is such that* $f(t) = |t|$, $t \in [-\varepsilon, \varepsilon]$ *for some* $\varepsilon > 0$ *and* f *is smooth on* $\mathbb{R} \setminus [-\varepsilon, \varepsilon]$. *Then, for a self-adjoint* $A \in L^p_{Haag}(\mathcal{M})$, *the restriction of* $\delta^A_{k,p,\phi}$ *to* $L^p_{Haag}(\mathcal{M})^{\times k}$ *is a symmetric multilinear bounded functional for every* $k = 1, \ldots, m$. *Moreover, for self-adjoint* $A, X \in L^p_{Haag}(\mathcal{M})$,

$$\|A + X\|^p_{L^p_{Haag}} - \|A\|^p_{L^p_{Haag}} - \sum_{k=1}^m \frac{1}{k} \delta^A_{k,p,\phi}\big(\underbrace{X, \ldots, X}_{k\text{-times}}\big) = O(\|X\|^p_{L^p_{Haag}}). \tag{5.8.4}$$

Proof It is proved in [170, Lemma 41] by utilizing Theorem 4.4.13 that $\delta^A_{k,p,\phi}$ is a symmetric multilinear bounded functional on $L^{p,\infty}(\mathcal{N}, \tau)^{\times k}$ for every $k = 1, \ldots, m$. Recalling that $L^p_{Haag}(\mathcal{M})$ is a closed linear subspace in $L^{p,\infty}(\mathcal{N}, \tau)$, we obtain that the restriction of $\delta^A_{k,p,\phi}$ to $L^p_{Haag}(\mathcal{M})^{\times k}$ (denoted again by $\delta^A_{k,p,\phi}$) is a symmetric multilinear bounded functional for every $k = 1, \ldots, m$.

We note that $(f^p)^{(m)} \in \Lambda_{p-m}$.

Applying the fundamental theorem of calculus to the function $t \mapsto \|A + tX\|^p_{L^p_{Haag}}$ gives

$$\|A + X\|^p_{L^p_{Haag}} - \|A\|^p_{L^p_{Haag}} = \int_0^1 \frac{d}{dt} \|A + tX\|^p_{L^p_{Haag}} \, dt. \tag{5.8.5}$$

By Lemma 5.8.4,

$$\frac{d}{dt} \|A + tX\|^p_{L^p_{Haag}} = \mathrm{tr}(X \cdot p|A + tX|^{p-1} \mathrm{sgn}(A + tX)). \tag{5.8.6}$$

Fix $t \in [0, 1]$. It is clear that $|A + tX|^{p-1}\mathrm{sgn}(A + tX) \in L^{\frac{p}{p-1}}_{Haag}(\mathcal{M})$ (see [206, Chapter II, Proposition 12]) and, therefore, $|A + tX|^{p-1}\mathrm{sgn}(A + tX) \cdot X \in L^1_{Haag}(\mathcal{M})$. By Lemma 2.9.4,

$$\mathrm{tr}(X \cdot p|A + tX|^{p-1}\mathrm{sgn}(A + tX)) = \phi(X \cdot p|A + tX|^{p-1}\mathrm{sgn}(A + tX)). \tag{5.8.7}$$

One can show that

$$X \cdot |A + tX|^{p-1} \text{sgn}(A + tX) - X \cdot (f^p)'(A + tX)$$

is a τ-finitely supported operator. Hence, using Lemmas 2.9.1 and 2.9.3, we infer

$$\phi(X \cdot p|A + tX|^{p-1}\text{sgn}(A + tX)) = \phi(X \cdot (f^p)'(A + tX)). \tag{5.8.8}$$

Combining (5.8.5)–(5.8.8) gives

$$\|A + X\|^p_{L^p_{Haag}} - \|A\|^p_{L^p_{Haag}} = \int_0^1 \phi(X \cdot (f^p)'(A + tX)) \, dt. \tag{5.8.9}$$

Next we claim that

$$(f^p)'(A + tX) = (f^p)'(A) + \sum_{k=1}^{m-1} t^k T^{A,\ldots,A}_{\varphi_{k,(f^p)(k+1)}}(X, \ldots, X)$$

$$+ t^{m-1}\left(T^{A+tX,A,\ldots,A}_{\varphi_{m-1,(f^p)(m)}}(X, \ldots, X) - T^{A,\ldots,A}_{\varphi_{m-1,(f^p)(m)}}(X, \ldots, X)\right). \tag{5.8.10}$$

Indeed, by (3.5.1) for $A, B \in (L^q + L^r)(\mathcal{M}, \tau)$, $1 < r < q < \infty$, and $f, f' \in C_b(\mathbb{R})$ (see Sect. 3.5)

$$(f^p)'(A + tX) - (f^p)'(A) = t T^{A+tX,A}_{\varphi_{1,(f^p)''}}(X).$$

It follows from Theorem 4.4.8 that

$$(f^p)'(A + tX) - (f^p)'(A) = t T^{A,A}_{\varphi_{1,(f^p)''}}(X) + t\left(T^{A+tX,A}_{\varphi_{1,(f^p)''}}(X) - T^{A,A}_{\varphi_{1,(f^p)''}}(X)\right)$$

$$= t T^{A,A}_{\varphi_{1,(f^p)''}}(X) + t^2 T^{A+tX,A,A}_{\varphi_{2,(f^p)(3)}}(X, X).$$

Repeating this process $m - 1$ times, we obtain (5.8.10).

Now we prove the Taylor expansion (5.8.4). Plugging (5.8.10) into (5.8.9), we obtain that

$$
\|A + X\|^p_{L^p_{Haag}} - \|A\|^p_{L^p_{Haag}}
$$

$$
= \phi(X \cdot (f^p)'(A)) + \sum_{k=1}^{m-1} \int_0^1 t^k \phi(X \cdot T^{A,\ldots,A}_{\varphi_{k,(f^p)^{(k+1)}}}(X,\ldots,X))dt
$$

$$
+ \int_0^1 t^{m-1} \cdot \phi\Big(X \cdot \Big(T^{A+tX,A,\ldots,A}_{\varphi_{m-1,(f^p)^{(m)}}}(X,\ldots,X) - T^{A,\ldots,A}_{\varphi_{m-1,(f^p)^{(m)}}}(X,\ldots,X)\Big)\Big)dt
$$

$$
= \phi(X \cdot (f^p)'(A)) + \sum_{k=1}^{m-1} \frac{1}{k+1}\phi(X \cdot T^{A,\ldots,A}_{\varphi_{k,(f^p)^{(k+1)}}}(X,\ldots,X))
$$

$$
+ \int_0^1 t^{m-1} \cdot \phi\Big(X \cdot \Big(T^{A+tX,A,\ldots,A}_{\varphi_{m-1,(f^p)^{(m)}}}(X,\ldots,X) - T^{A,\ldots,A}_{\varphi_{m-1,(f^p)^{(m)}}}(X,\ldots,X)\Big)\Big)dt.
$$

By (5.8.3),

$$
\|A + X\|^p_{L^p_{Haag}} - \|A\|^p_{L^p_{Haag}} - \sum_{k=1}^{m} \frac{1}{k}\delta^A_{k,p,\phi}\Big(\underbrace{X,\ldots,X}_{k\text{-times}}\Big)
$$

$$
= \int_0^1 t^{m-1} \cdot \phi\Big(X \cdot \Big(T^{A+tX,A,\ldots,A}_{\varphi_{m-1,(f^p)^{(m)}}}(X,\ldots,X) - T^{A,\ldots,A}_{\varphi_{m-1,(f^p)^{(m)}}}(X,\ldots,X)\Big)\Big)dt.
$$

$$\tag{5.8.11}$$

Next, we estimate the integral on the right hand side of (5.8.11). Using Lemmas 2.9.2 and 2.6.1(i) and Theorem 4.4.11 with $k = m - 1$, $\alpha = p - m$, $h = (f^p)^{(m)}$, $A = A + tX$, $B = A$, we obtain

$$
\Big|\phi\Big(X \cdot \Big(T^{A+tX,\ldots,A}_{\varphi_{m-1,(f^p)^{(m)}}}(X,\ldots,X) - T^{A,\ldots,A}_{\varphi_{m-1,(f^p)^{(m)}}}(X,\ldots,X)\Big)\Big)\Big|
$$

$$
\leqslant 2\|\phi\|\|X\|_{L^{p,\infty}}\Big\|T^{A+tX,\ldots,A}_{\varphi_{m-1,(f^p)^{(m)}}}(X,\ldots,X) - T^{A,\ldots,A}_{\varphi_{m-1,(f^p)^{(m)}}}(X,\ldots,X)\Big\|_{L^{\frac{p}{p-1},\infty}}
$$

$$
\leqslant 2C(m-1,p)\|\phi\|\|X\|_{L^{p,\infty}}\|(f^p)^{(m)}\|_{\Lambda_{p-m}}\|tX\|^{p-m}_{L^{p,\infty}}\|X\|^{m-1}_{L^{p,\infty}}
$$

$$
= 2C(m-1,p)\,t^{p-m}\|\phi\|\|(f^p)^{(m)}\|_{\Lambda_{p-m}}\|X\|^p_{L^{p,\infty}}.
$$

Therefore, by (2.6.1) and (2.7.4),

$$\left| \int_0^1 t^{m-1} \cdot \phi\Big(X \cdot \Big(T_{\varphi_{m-1,(f^p)^{(m)}}}^{A+tX,A,\dots,A}(X,\dots,X) - T_{\varphi_{m-1,(f^p)^{(m)}}}^{A,\dots,A}(X,\dots,X) \Big) \Big) dt \right|$$

$$\leqslant 2C(m-1,p)\|\phi\|\|(f^p)^{(m)}\|_{\Lambda_{p-m}}\|X\|_{L^{p,\infty}}^p \int_0^1 t^{p-1}dt$$

$$\leqslant \operatorname{const}\|X\|_{L_{Haag}^p}^p,$$

where the constant is independent of A, X. This completes the proof of (5.8.4). □

We now complete the proof of Theorem 5.8.1.

Proof (Proof of Theorem 5.8.1) It suffices to verify the assertion of Theorem 5.8.1 only for self-adjoint elements. Indeed, for all self-adjoint A, $X \in L_{Haag}^p(\mathcal{M})$ the existence of $\delta_{k,p,\phi}^A$ satisfying (5.8.4) is established in Theorem 5.8.5. For an arbitrary $X \in L_{Haag}^p(\mathcal{M})$ consider the mapping

$$\alpha(X) = \frac{1}{2^{1/p}}\begin{pmatrix} 0 & X \\ X^* & 0 \end{pmatrix}.$$

Observing that $L_{Haag}^p(\mathcal{M}\overline{\otimes}M_2) = L_{Haag}^p(\mathcal{M})\overline{\otimes}L^p(M_2,\mathrm{Tr})$, where M_2 is the set of 2×2 matrices (see, e.g., [94, p. 70]), we conclude that α is an isometric embedding of $L_{Haag}^p(\mathcal{M})$ into the real subspace of all self-adjoint operators from $L_{Haag}^p(\mathcal{M}\overline{\otimes}M_2)$. For arbitrary operators $A, X_1,\dots,X_k \in L_{Haag}^p(\mathcal{M})$, we set

$$\delta_{k,p,\phi}^A(X_1,\dots,X_k) := \frac{1}{2}\delta_{k,p,\phi}^{\alpha(A)}(\alpha(X_1),\dots,\alpha(X_k)), \quad k=1,\dots,m.$$

It is straightforward that $\delta_{k,p,\phi}^A$ is a bounded symmetric multilinear form satisfying (5.8.4).

Application of Proposition 5.8.3, Lemma 5.8.4, and Theorem 5.8.5 completes the proof.   □

References

1. A. Abdessemed, E.B. Davies, Some commutator estimates in the Schatten classes. J. Lond. Math. Soc. **39**, 299–308 (1989)
2. V.M. Adamjan, H. Neidhardt, On the summability of the spectral shift function for pair of contractions and dissipative operators. J. Oper. Theory **24**(1), 187–205 (1990)
3. V.M. Adamjan, B.S. Pavlov, Trace formula for dissipative operators. Vestn. Leningr. Univ. Mat. Mekh. Astronom. **7**(2), 5–9 (1979, in Russian)
4. A.B. Aleksandrov, V.V. Peller, Functions of operators under perturbations of class $\mathcal{S}^p$. J. Funct. Anal. **258**(11), 3675–3724 (2010)
5. A.B. Aleksandrov, V.V. Peller, Operator Hölder-Zygmund functions. Adv. Math. **224**(3), 910–966 (2010)
6. A.B. Aleksandrov, V.V. Peller, Functions of perturbed dissipative operators. Algebra i Analiz **23**(2), 9–51 (2011). Translation: St. Petersburg Math. J. **23**(2), 209–238 (2012)
7. A.B. Aleksandrov, V.V. Peller, Trace formulae for perturbations of class S_m. J. Spectral Theory **1**(1), 1–26 (2011)
8. A.B. Aleksandrov, V.V. Peller, Estimates of operator moduli of continuity. J. Funct. Anal. **261**, 2741–2796 (2011)
9. A.B. Aleksandrov, V.V. Peller, Operator Lipschitz functions. Uspekhi Mat. Nauk **71**(4(430)), 3–106 (2016, in Russian). Translation: Russ. Math. Surv. **71**(4), 605–702 (2016)
10. A.B. Aleksandrov, V.V. Peller, Krein's trace formula for unitary operators and operator Lipschitz functions. Funct. Anal. Appl. **50**(3) 167–175 (2016)
11. A.B. Aleksandrov, V.V. Peller, Multiple operator integrals, Haagerup and Haagerup-like tensor products, and operator ideals. Bull. Lond. Math. Soc. **49**, 463–479 (2017)
12. A.B. Aleksandrov, V.V. Peller, Dissipative operators and operator Lipschitz functions. Proc. Am. Math. Soc. **147**(5), 2081–2093 (2019)
13. A.B. Aleksandrov, V.V. Peller, D. Potapov, F.A. Sukochev, Functions of normal operators under perturbations. Adv. Math. **226**(6), 5216–5251 (2011)
14. A.B. Aleksandrov, F.L. Nazarov, V.V. Peller, Functions of noncommuting self-adjoint operators under perturbation and estimates of triple operator integrals. Adv. Math. **295**, 1–52 (2016)
15. H. Araki, T. Masuda, Positive cones and Lp-spaces for von Neumann algebras. Publ. Res. Inst. Math. Sci. **18**(2), 759–831 (339–411) (1982)
16. J. Arazy, Some remarks on interpolation theorems and the boundness of the triangular projection in unitary matrix spaces. Integr. Equ. Oper. Theory **1**, 453–495 (1978)
17. J. Arazy, Certain Schur-Hadamard multipliers in the space C_p. Proc. Am. Math. Soc. **86**(1), 59–64 (1982)

© Springer Nature Switzerland AG 2019

A. Skripka, A. Tomskova, *Multilinear Operator Integrals*,
Lecture Notes in Mathematics 2250, https://doi.org/10.1007/978-3-030-32406-3

18. J. Arazy, L. Zelenko, Directional operator differentiability of non-smooth functions. J. Oper. Theory **55**(1), 49–90 (2006)

19. J. Arazy, T.J. Barton, Y. Friedman, Operator differentiable functions. Integr. Equ. Oper. Theory **13**(4), 462–487 (1990)

20. M.F. Atiyah, V.K. Patodi, I.M. Singer, Spectral asymmetry and Riemannian geometry. III. Math. Proc. Camb. Philos. Soc. **79**(1), 71–99 (1976)

21. P.J. Ayre, M.G. Cowling, F.A. Sukochev, Operator Lipschitz estimates in the unitary setting. Proc. Am. Math. Soc. **144**(3), 1053–1057 (2016)

22. N.A. Azamov, P.G. Dodds, F.A. Sukochev, The Krein spectral shift function in semifinite von Neumann algebras. Integr. Equ. Oper. Theory **55**, 347–362 (2006)

23. N.A. Azamov, A.L. Carey, F.A. Sukochev, The spectral shift function and spectral flow. Commun. Math. Phys. **276**(1), 51–91 (2007)

24. N.A. Azamov, A.L. Carey, P.G. Dodds, F.A. Sukochev, Operator integrals, spectral shift, and spectral flow. Can. J. Math. **61**(2), 241–263 (2009)

25. M.-T. Benameur, A.L. Carey, J. Phillips, A. Rennie, F.A. Sukochev, K.P. Wojciechowski, An analytic approach to spectral flow in von Neumann algebras, in *Analysis, Geometry and Topology of Elliptic Operators* (World Scientific Publishing, Hackensack, 2006), pp. 297–352

26. G. Bennett, Unconditional convergence and almost everywhere convergence. Z. Wahrsch. Verw. Geb. **34**(2), 135–155 (1976)

27. G. Bennett, Schur multipliers. Duke Math. J. **44**(3), 603–639 (1977)

28. C. Bennett, R. Sharpley, *Interpolation of Operators*. Pure and Applied Mathematics, vol. 129 (Academic, Boston, 1988)

29. P.H. Bérard, *Spectral Geometry: Direct and Inverse Problems*. Lecture Notes in Mathematics, vol. 1207 (Springer, Berlin, 1986)

30. J. Bergh, J. Löfström, *Interpolation Spaces. An Introduction*. Grundlehren der Mathematischen Wissenschaften, vol. 223 (Springer, Berlin, 1976)

31. A.T. Bernardino, A simple natural approach to the uniform boundedness principle for multilinear mappings. Proyecciones J. Math. **28**(3), 203–207 (2009)

32. R. Bhatia, *Matrix Analysis*. Graduate Texts in Mathematics, vol. 169 (Springer, New York, 1997)

33. M.Sh. Birman, A.B. Pushnitski, Spectral shift function, amazing and multifaceted. Dedicated to the memory of Mark Grigorievich Krein (1907–1989). Integr. Equ. Oper. Theory **30**(2), 191–199 (1998)

34. M.S. Birman, M.Z. Solomyak, *Double Stieltjes Operator Integrals*. Problems of Mathematical Physics, Izdat (Leningrad University, Leningrad, 1966, in Russian), pp. 33–67. English translation in: Topics in Mathematical Physics, vol. 1 (Consultants Bureau Plenum Publishing Corporation, New York, 1967), pp. 25–54

35. M.S. Birman, M.Z. Solomyak, *Double Stieltjes Operator Integrals II*. Problems of Mathematical Physics, Izdat, no. 2 (Leningrad University, Leningrad, 1967, in Russian), pp. 26–60. English translation in: Topics in Mathematical Physics, vol. 2 (Consultants Bureau, New York, 1968), pp. 19–46

36. M.S. Birman, M.Z. Solomyak, *Double Stieltjes Operator Integrals III*. Problems of Mathematical Physics, vol. 6 (Leningrad University, Leningrad, 1973, in Russian), pp. 27–53

37. M.S. Birman, M.Z. Solomyak, Remarks on the spectral shift function. J. Sov. Math. **3**, 408–419 (1975)

38. M.S. Birman, M.Z. Solomyak, *Spectral Theory of Self-adjoint Operators in Hilbert Space* (D. Reidel Publishing Company, Netherlands, 1987)

39. M.S. Birman, M.Z. Solomyak, Operator integration, perturbations and commutators. Zap. Nauchn. Sem. Leningrad. Otdel. Mat. Inst. Steklov. (LOMI) **170**, 34–66 (1989, in Russian), Issled. Linein. Oper. Teorii Funktsii. **17**, 321. Translation: J. Sov. Math. **63**(2), 129–148 (1993)

40. M.S. Birman, M.Z. Solomyak, Tensor product of a finite number of spectral measures is always a spectral measure. Integr. Equ. Oper. Theory **24**(2), 179–187 (1996)

41. M.S. Birman, M.Z. Solomyak, Double operator integrals in a Hilbert space. Integr. Equ. Oper. Theory **47**(2), 131–168 (2003)

42. M.S. Birman, D.R. Yafaev, The spectral shift function. The papers of M. G. Krein and their further development. Algebra i Analiz **4**(5), 1–44 (1992, in Russian). Translation: St. Petersburg Math. J. **4**(5), 833–870 (1993)

43. M.S. Birman, M.Z. Solomyak, A.M. Vershik, The product of commuting spectral measures may fail to be countably additive. Funktsional. Anal. i Prilozhen. **13**(1), 61–62 (1979, in Russian)

44. R. Bonic, J. Frampton, Smooth functions on Banach manifolds. J. Math. Mech. **15**(5), 877–898 (1966)

45. A. Carey, J. Phillips, Unbounded Fredholm modules and spectral flow. Can. J. Math. **50**(4), 673–718 (1998)

46. A. Carey, J. Phillips, Spectral flow in Fredholm modules, eta invariants and the JLO cocycle. K-Theory **31**(2), 135–194 (2004)

47. R.W. Carey, J.D. Pincus, Mosaics, principal functions, and mean motion in von Neumann algebras. Acta Math. **138**, 153–218 (1977)

48. M. Caspers, S. Montgomery-Smith, D. Potapov, F. Sukochev, The best constants for operator Lipschitz functions on Schatten classes. J. Funct. Anal. **267**(10), 3557–3579 (2014)

49. M. Caspers, D. Potapov, F. Sukochev, D. Zanin, Weak type commutator and Lipschitz estimates: resolution of the Nazarov-Peller conjecture. Am. J. Math. **141**(3), 593–610 (2019)

50. A. Chattopadhyay, A. Skripka, Trace formulas for relative Schatten class perturbations. J. Funct. Anal. **274**, 3377–3410 (2018)

51. V.I. Chilin, F.A. Sukochev, Weak convergence in non-commutative symmetric spaces. J. Oper. Theory **31**(1), 35–65 (1994)

52. C. Coine, Perturbation theory and higher order S^p-differentiability of operator functions (2019). arXiv:1906.05585

53. C. Coine, C. Le Merdy, D. Potapov, F. Sukochev, A. Tomskova, Resolution of Peller's problem concerning Koplienko-Neidhardt trace formula. Proc. Lond. Math. Soc. (3) **113**(2), 113–139 (2016)

54. C. Coine, C. Le Merdy, D. Potapov, F. Sukochev, A. Tomskova, Peller's problem concerning Koplienko-Neidhardt trace formula: the unitary case. J. Funct. Anal. **271**(7), 1747–1763 (2016)

55. C. Coine, C. Le Merdy, A. Skripka, F. Sukochev, Higher order S^2-differentiability and application to Koplienko trace formula. J. Funct. Anal. **276**(10), 3170–3204 (2019)

56. C. Coine, C. Le Merdy, F. Sukochev, When do triple operator integrals take value in the trace class? (2019). arXiv:1706.01662

57. A. Connes, *Noncommutative Geometry* (Academic, San Diego, 1994)

58. A. Connes, D. Sullivan, N. Teleman, Quasiconformal mappings, operators on Hilbert space, and local formulae for characteristic classes. Topology **33**(4), 663–681 (1994)

59. J.H. Curtiss, Limits and bounds for divided differences on a Jordan curve in the complex domain. Pac. J. Math. **12**, 1217–1233 (1962)

60. H.G. Dales, *Banach Algebras and Automatic Continuity*. Mathematical Society Monographs (Oxford University Press, London, 2000)

61. Yu.L. Daletskii, S.G. Krein, Integration and differentiation of functions of Hermitian operators and application to the theory of perturbations. Trudy Sem. Functsion. Anal. Voronezh. Gos. Univ. **1**, 81–105 (1956, in Russian)

62. E.B. Davies, Lipschitz continuity of functions of operators in the Schatten classes. J. Lond. Math. Soc. **37**, 148–157 (1988)

63. B. de Pagter, F. Sukochev, Differentiation of operator functions in non-commutative L_p-spaces. J. Funct. Anal. **212**(1), 28–75 (2004)

64. B. de Pagter, H. Witvliet, F.A. Sukochev, Double operator integrals. J. Funct. Anal. **192**(1), 52–111 (2002)

65. R.A. DeVore, G.G. Lorentz, *Constructive Approximation*. Grundlehren der Mathematischen Wissenschaften [Fundamental Principles of Mathematical Sciences], vol. 303 (Springer, Berlin, 1993)

66. J. Diestel, J.J. Uhl, *Vector Measures*. Mathematical Surveys, no. 15 (American Mathematical Society, Providence, 1977)

67. J. Diestel, H. Jarchow, A. Tonge, *Absolutely Summing Operators*. Cambridge Studies in Advanced Mathematics, vol. 43 (Cambridge University Press, Cambridge, 1995)

68. S.J. Dilworth, *Special Banach Lattices and Their Applications*. Handbook of the Geometry of Banach Spaces, vol. I (North-Holland, Amsterdam, 2001), pp. 497–532

69. P.G. Dodds, T.K. Dodds, B. de Pagter, Fully symmetric operator spaces. Integr. Equ. Oper. Theory **15**(6), 942–972 (1992)

70. P.G. Dodds, T.K. Dodds, B. de Pagter, F.A. Sukochev, Lipschitz continuity of the absolute value and Riesz projections in symmetric operator spaces. J. Funct. Anal. **148**(1), 28–69 (1997)

71. P.G. Dodds, T.K. Dodds, F.A. Sukochev, D. Zanin, Arithmetic-geometric mean and related submajorisation and norm inequalities for τ-measurable operators, preprint

72. N. Dunford, J.T. Schwartz. *Linear Operators. Part III: Spectral Operators* (Interscience Publishers [Wiley], New York, 1971). With the assistance of W. G. Bade and R. G. Bartle, Pure and Applied Mathematics, Vol. VII

73. K. Dykema, N.J. Kalton, Sums of commutators in ideals and modules of type II factors. Ann. Inst. Fourier (Grenoble) **55**(3), 931–971 (2005)

74. K. Dykema, A. Skripka, Higher order spectral shift. J. Funct. Anal. **257**, 1092–1132 (2009)

75. K. Dykema, A. Skripka, Perturbation formulas for traces on normed ideals. Commun. Math. Phys. **325**(3), 1107–1138 (2014)

76. K. Dykema, A. Skripka, Erratum to: Perturbation formulas for traces on normed ideals. Commun. Math. Phys. **340**(2), 865 (2015)

77. K. Dykema, A. Skripka, Hölder's inequality for roots of symmetric operator spaces. Stud. Math. **228**(1), 47–54 (2015)

78. E.G. Effros, Z.-J. Ruan, *Multivariable Multipliers for Groups and Their Operator Algebras*. Proceedings of Symposia in Pure Mathematics, vol. 51, Part 1 (American Mathematical Society, Providence, 1990), pp. 197–218

79. T. Fack, H. Kosaki, Generalized s-numbers of τ-measurable operators. Pac. J. Math. **123**, 269–300 (1986)

80. Yu.B. Farforovskaya, An example of a Lipschitz function of selfadjoint operators that yields a non-nuclear increase under a nuclear perturbation. J. Sov. Math. **4**, 426–433 (1975, in Russian)

81. Yu.B. Farforovskaja, An estimate of the norm of $|f(B) - f(A)|$ for selfadjoint operators A and B, Investigations on linear operators and theory of functions, VI. Zap. Nauchn. Sem. Leningrad. Otdel. Mat. Inst. Steklov. (LOMI) **56**, 143–162 (1976). J. Soviet Math. **14**(2), 1133–1149 (1980, in Russian)

82. C.K. Fong, G.J. Murphy, Ideals and Lie ideals of operators. Acta Sci. Math. **51**, 441–456 (1987)

83. F. Gesztesy, A. Pushnitski, B. Simon, On the Koplienko spectral shift function. I. Basics. Zh. Mat. Fiz. Anal. Geom. **4**(1), 63–107 (2008)

84. E. Getzler, The odd Chern character in cyclic homology and spectral flow. Topology **32**(3), 489–507 (1993)

85. I.C. Gohberg, M.G. Krein, *Introduction to the Theory of Linear Nonselfadjoint Operators*. Translations of Mathematical Monographs, vol. 18 (American Mathematical Society, Providence, 1969)

86. D. Guido, T. Isola, Singular traces on semifinite von Neumann algebras. J. Funct. Anal. **134**(2), 451–485 (1995)

87. U. Haagerup, L_p-spaces associated with an arbitrary von Neumann algebra, in *Algèbres d'opérateurs et leurs applications en physique mathématique* (Proc. Colloq., Marseille, 1977), pp. 175–184, Colloques Internationaux du CNRS, 274, CNRS, Paris, 1979

88. F. Hiai, Matrix analysis: matrix monotone functions, matrix means and majorization. Interdiscip. Inf. Sci. **16**(2), 139–248 (2010)

89. F. Hiai, H. Kosaki, *Means of Hilbert Space Operators*. Lecture Notes in Mathematics, vol. 1820 (Springer, Berlin, 2003)

90. R.A. Horn, C.R. Johnson, *Matrix Analysis*, 2nd edn. (Cambridge University Press, Cambridge, 2013)
91. T. Hytönen, J. van Neerven, M. Veraar, L. Weis, *Analysis in Banach Spaces*. Vol. I. Martingales and Littlewood-Paley Theory. Ergebnisse der Mathematik und ihrer Grenzgebiete 3. Folge. A Series of Modern Surveys in Mathematics, vol. 63 (Springer, Cham, 2016)
92. S. Janson, T.H. Wolff, Schatten classes and commutators of singular integral operators. Ark. Mat. **20**(2), 301–310 (1982)
93. B.E. Johnson, J.P. Williams, The range of normal derivations. Pac. J. Math. **58**, 105–122 (1975)
94. M. Junge, Z-J. Ruan, Q. Xu, Rigid $\mathcal{OL}_p$ structures of non-commutative L_p-spaces associated with hyperfinite von Neumann algebras. Math. Scand. **96**, 63–95 (2005)
95. K. Juschenko, I.G. Todorov, L. Turowska, Multidimensional operator multipliers. Trans. Am. Math. Soc. **361**, 4683–4720 (2009)
96. R.V. Kadison, J.R. Ringrose, *Fundamentals of the Theory of Operator Algebras. Vol. II. Advanced Theory*. Pure and Applied Mathematics, vol. 100 (Academic, Orlando, 1986), XIV, 1074 pp.
97. N.J. Kalton, F. Sukochev, Symmetric norms and spaces of operators. J. Reine Angew. Math. **621**, 81–121 (2008)
98. N.J. Kalton, N.T. Peck, J.W. Roberts, *An F-space Sampler*. London Mathematical Society Lecture Note Series, vol. 89 (Cambridge University Press, Cambridge, 1984)
99. T. Kato, Continuity of the map $S \mapsto |S|$ for linear operators. Proc. Jpn. Acad. **49**, 157–160 (1973)
100. T. Kato, *Perturbation Theory for Linear Operators*. Classics in Mathematics (Springer, Berlin, 1995)
101. E. Kissin, V.S. Shulman, Operator-differentiable functions and derivations of operator algebras. Funktsional. Anal. i Prilozhen. **30**(4), 75–77 (1996, in Russian). Translation: Funct. Anal. Appl. **30**(4), 280–282 (1996)
102. E. Kissin, V.S. Shulman, On the range inclusion of normal derivations: variations on a theme by Johnson, Williams and Fong. Proc. Lond. Math. Soc. **83**, 176–198 (2001)
103. E. Kissin, V.S. Shulman, Classes of operator-smooth functions. II. Operator-differentiable functions. Integr. Equ. Oper. Theory **49**(2), 165–210 (2004)
104. E. Kissin, V.S. Shulman, Classes of operator-smooth functions. I. Operator-Lipschitz functions. Proc. Edinb. Math. Soc. **48**, 151–173 (2005)
105. E. Kissin, V.S. Shulman, Classes of operator-smooth functions. III Stable functions and Fuglede ideals. Proc. Edinb. Math. Soc. **48**, 175–197 (2005)
106. E. Kissin, V.S. Shulman, Lipschitz functions on Hermitian Banach *-algebras. Q. J. Math. **57**, 215–239 (2006)
107. E. Kissin, V.S. Shulman, On fully operator Lipschitz functions. J. Funct. Anal. **253**(2), 711–728 (2007)
108. E. Kissin, V.S. Shulman, Functions acting on symmetrically normed ideals and on the domains of derivations on these ideals. J. Oper. Theory **57**(1), 63–82 (2007)
109. E. Kissin, V.S. Shulman, L.B. Turowska, Extension of operator Lipschitz and commutator bounded functions. Oper. Theory Adv. Appl. **171**, 225–244 (2006)
110. E. Kissin, D. Potapov, V. Shulman, F. Sukochev, Operator smoothness in Schatten norms for functions of several variables: Lipschitz conditions, differentiability and unbounded derivations. Proc. Lond. Math. Soc. **108**(3), 327–349 (2014)
111. L.S. Koplienko, Trace formula for perturbations of nonnuclear type. Sibirsk. Mat. Zh. **25**, 62–71 (1984, in Russian). Translation: Siberian Math. J. **25**, 735–743 (1984)
112. H. Kosaki, Applications of uniform convexity of noncommutative L_p-spaces. Trans. Am. Math. Soc. **283**(1), 265–282 (1984)
113. M.G. Krein, On a trace formula in perturbation theory. Mat. Sbornik **33**, 597–626 (1953, in Russian)
114. M.G. Krein, On the perturbation determinant and the trace formula for unitary and self-adjoint operators. Dokl. Akad. Nauk SSSR **144**, 268–271 (1962, in Russian). Translation: Soviet Math. Dokl. **3**, 707–710 (1962)

115. M.G. Krein, Some new studies in the theory of perturbations of self-adjoint operators. 1964 First Mathematical Summer School, Part I, pp. 103–187 Izdat. "Naukova Dumka", Kiev (1964, in Russian)

116. M.G. Krein, On perturbation determinants and a trace formula for certain classes of pairs of operators. J. Oper. Theory **17**(1), 129–187 (1987, in Russian). Translation: Am. Math. Soc. Trans. **145**(2), 39–84 (1989)

117. S. Krein, Ju. Petunin, E. Semenov, *Interpolation of Linear Operators* (Nauka, Moscow, 1978, in Russian). Translation: Translations of Mathematical Monographs, vol. 54 (American Mathematical Society, Providence, 1982)

118. H. Langer, Eine Erweiterung der Spurformel der Stoërungstheorie. Math. Nachr. **30**, 123–135 (1965, in German)

119. C. Le Merdy, A. Skripka, Higher order differentiability of operator functions in Schatten norms. J. Inst. Math. Jussieu. https://doi.org/10.1017/S1474748019000033

120. I.M. Lifshits, On a problem of the theory of perturbations connected with quantum statistics. Uspehi Mat. Nauk (N.S.) **7**(1(47)), 171–180 (1952, in Russian)

121. J. Lindenstrauss, L. Tzafriri, *Classical Banach Spaces I: Sequence Spaces*. Ergebnisse der Mathematik und ihrer Grenzgebiete, vol. 92 (Springer, Berlin, 1977)

122. J. Lindenstrauss, L. Tzafriri, *Classical Banach Spaces II: Function Spaces*. Ergebnisse der Mathematik und ihrer Grenzgebiete, vol. 97 (Springer, Berlin, 1979)

123. S. Lord, F. Sukochev, D. Zanin, *Singular Traces: Theory and Applications*. Studies in Mathematics, vol. 46 (De Gruyter, Berlin, 2012)

124. S. Lord, E. McDonald, F. Sukochev, D. Zanin, Quantum differentiability of essentially bounded functions on Euclidean space. J. Funct. Anal. **273**(7), 2353–2387 (2017)

125. K. Löwner, Über monotone Matrixfunktionen. Math. Z. **38**(1), 177–216 (1934, in German)

126. L.A. Lusternik, V.J. Sobolev, *Elements of Functional Analysis* (Hindustan Publishing Corporation/Halsted Press, Delhi/New York, 1974)

127. K.A. Makarov, A. Skripka, M. Zinchenko, On perturbation determinant for antidissipative operators. Integr. Equ. Oper. Theory **81**(3), 301–317 (2015)

128. M.M. Malamud, H. Neidhardt, Trace formulas for additive and non-additive perturbations. Adv. Math. **274**, 736–832 (2015)

129. M.M. Malamud, H. Neidhardt, V.V. Peller, Analytic operator Lipschitz functions in the disk and a trace formula for functions of contractions. Funct. Anal. Appl. **51**(3), 33–55 (2017)

130. M.M. Malamud, H. Neidhardt, V.V. Peller, Absolute continuity of spectral shift. J. Funct. Anal. **276**(5), 1575–1621 (2019)

131. A. McIntosh, Counterexample to a question on commutators. Proc. Am. Math. Soc. **29**, 337–340 (1971)

132. B.Sz. Nagy, C. Foias, *Harmonic Analysis of Operators on Hilbert Space* (North-Holland/American Elsevier/Akade'miai Kiado', Amsterdam/New York/Budapest, 1970)

133. F. Nazarov, V. Peller, Lipschitz functions of perturbed operators. C. R. Math. Acad. Sci. Paris **347**(15–16), 857–862 (2009)

134. H. Neidhardt, Scattering matrix and spectral shift of the nuclear dissipative scattering theory, in *Operators in Indefinite Metric Spaces, Scattering Theory and Other Topics* (Bucharest, 1985). Operator Theory Advances and Applications, vol. 24 (Birkhäuser, Basel, 1987), pp. 237–250

135. H. Neidhardt, Scattering matrix and spectral shift of the nuclear dissipative scattering theory II. J. Oper. Theory **19**(1), 43–62 (1988)

136. H. Neidhardt, Spectral shift function and Hilbert-Schmidt perturbation: extensions of some work of L.S. Koplienko. Math. Nachr. **138**, 7–25 (1988)

137. J. Parcet, Pseudo-localization of singular integrals and noncommutative Calderón-Zygmund theory. J. Funct. Anal. **256**(2), 509–593 (2009)

138. B.S. Pavlov, Multidimensional operator integrals, in *Linear Operators and Operator Equations*. Problems in Mathematical Analysis, no. 2 (Izdat Leningrad University, Leningrad, 1969, in Russian), pp. 99–122

139. G.K. Pedersen, Operator differentiable functions. Publ. Res. Inst. Math. Sci. **36**(1), 139–157 (2000)
140. J. Peetre, *New Thoughts on Besov Spaces*. Duke University Mathematics Series (Duke University, Durham, 1976)
141. V.V. Peller, Hankel operators in the theory of perturbations of unitary and selfadjoint operators. Funktsional. Anal. i Prilozhen. **19**(2), 37–51 (1985, in Russian). Translation: Funct. Anal. Appl. **19**, 111–123 (1985)
142. V.V. Peller, *For Which f does $A - B \in S_p$ Imply that $f(A) - f(B) \in S_p$?* Operator Theory: Advances and Applications, vol. 24, (Birkhäuser, Basel, 1987), pp. 289–294
143. V.V. Peller, Hankel operators in the perturbation theory of unbounded selfadjoint operators, in *Analysis and Partial Differential Equations*. Lecture Notes in Pure and Applied Mathematics, vol. 122 (Dekker, New York, 1990), pp. 529–544
144. V.V. Peller, *Hankel Operators and their Applications*. Springer Monographs in Mathematics (Springer, New York, 2003)
145. V.V. Peller, An extension of the Koplienko-Neidhardt trace formulae. J. Funct. Anal. **221**, 456–481 (2005)
146. V.V. Peller, Multiple operator integrals and higher operator derivatives. J. Funct. Anal. **233**(2), 515–544 (2006)
147. V.V. Peller, Differentiability of functions of contractions, in *Linear and Complex Analysis*. American Mathematical Society Translations: Series 2, vol. 226, Advances in the Mathematical Sciences, vol. 63 (American Mathematical Society, Providence, 2009), pp. 109–131
148. V.V. Peller, The Lifshits-Krein trace formula and operator Lipschitz functions. Proc. Am. Math. Soc. **144**(12), 5207–5215 (2016)
149. V.V. Peller, Multiple operator integrals in perturbation theory. Bull. Math. Sci. **6**, 15–88 (2016)
150. J. Phillips, Self-adjoint Fredholm operators and spectral flow. Can. Math. Bull. **39**(4), 460–467 (1996)
151. J. Phillips, Spectral flow in type I and II factors – a new approach. Fields Inst. Commun. **17**, 137–153 (1997). American Mathematical Society, Providence
152. A. Pietsch, *Operator Ideals*. North-Holland Mathematical Library, vol. 20 (North-Holland, Amsterdam, 1980)
153. G. Pisier, *Similarity Problems and Completely Bounded Maps*. Lecture Notes in Mathematics, vol. 1618 (Springer, Berlin, 2001)
154. G. Pisier, Q. Xu, *Non-commutative L_p-spaces*. Handbook of the Geometry of Banach Spaces, vol. 2 (North-Holland, Amsterdam, 2003), pp. 1459–1517
155. D. Potapov, Lipschitz and commutator estimates, a unified approach, Dissertation, Ph.D. Mathematics, Flinders University, 2007
156. D. Potapov, F. Sukochev, Lipschitz and commutator estimates in symmetric operator spaces. J. Oper. Theory **59**(1), 211–234 (2008)
157. D. Potapov, F. Sukochev, Unbounded Fredholm modules and double operator integrals. J. Reine Angew. Math. **626**, 159–185 (2009)
158. D. Potapov, F. Sukochev, Double operator integrals and submajorization. Math. Model. Nat. Phenom. **5**(4), 317–339 (2010)
159. D. Potapov, F. Sukochev, Operator-Lipschitz functions in Schatten-von Neumann classes. Acta Math. **207**(2), 375–389 (2011)
160. D. Potapov, F. Sukochev, Koplienko spectral shift function on the unit circle. Commun. Math. Phys. **309**, 693–702 (2012)
161. D. Potapov, F. Sukochev, Fréchet differentiability of S^p norms. Adv. Math. **262**, 436–475 (2014)
162. D. Potapov, A. Skripka, F. Sukochev, On Hilbert-Schmidt compatibility. Oper. Matrices **7**(1), 1–34 (2013)
163. D. Potapov, A. Skripka, F. Sukochev, Spectral shift function of higher order. Invent. Math. **193**(3), 501–538 (2013)
164. D. Potapov, A. Skripka, F. Sukochev, Spectral shift function of higher order for contractions. Proc. Lond. Math. Soc. **108**(3), 327–349 (2014)

165. D. Potapov, F. Sukochev, D. Zanin, Krein's trace theorem revisited. J. Spectr. Theory **4**(2), 415–430 (2014)

166. D. Potapov, A. Skripka, F. Sukochev, Trace formulas for resolvent comparable operators. Adv. Math. **272**, 630–651 (2015)

167. D. Potapov, F. Sukochev, A. Usachev, D. Zanin, Singular traces and perturbation formulae of higher order. J. Funct. Anal. **269**(5), 1441–1481 (2015)

168. D. Potapov, A. Skripka, F. Sukochev, Functions of unitary operators: derivatives and trace formulas. J. Funct. Anal. **270**(6), 2048–2072 (2016)

169. D. Potapov, A. Skripka, F. Sukochev, A. Tomskova, Multilinear Schur multipliers and applications to operator Taylor remainders. Adv. Math. **320**, 1063–1098 (2017)

170. D. Potapov, F. Sukochev, A. Tomskova, D. Zanin, Fréchet differentiability of the norm of L_p-spaces associated with arbitrary von Neumann algebras. Trans. Am. Math. Soc. **371**(11), 7493–7532 (2019)

171. M. Reed, B. Simon, *Methods of Modern Mathematical Physics. I. Functional Analysis*, 2nd edn. (Academic, New York, 1980)

172. R. Rochberg, S. Semmes, Nearly weakly orthonormal sequences, singular value estimates, and Calderon-Zygmund operators. J. Funct. Anal. **86**(2), 237–306 (1989)

173. J. Rozendaal, F. Sukochev, A. Tomskova, Operator Lipschitz functions on Banach spaces. Stud. Math. **232**(1), 57–92 (2016)

174. R. Ryan, *Introduction to Tensor Products of Banach Spaces*. Springer Monographs in Mathematics (Springer, London, 2002)

175. A.V. Rybkin, The spectral shift function for a dissipative and a selfadjoint operator, and trace formulas for resonances. Mat. Sb. (N.S.) **125(167)**(3), 420–430 (1984, in Russian)

176. A.V. Rybkin, A trace formula for a contractive and a unitary operator. Funktsional. Anal. i Prilozhen. **21**(4), 85–87 (1987, in Russian)

177. A.V. Rybkin, The discrete and the singular spectrum in the trace formula for a contractive and a unitary operator. Funktsional. Anal. i Prilozhen. **23**(3), 84–85 (1989, in Russian). Translation: Funct. Anal. Appl. **23**(3), 244–246 (1990)

178. A.V. Rybkin, The spectral shift function, the characteristic function of a contraction and a generalized integral. Mat. Sb. **185**(10), 91–144 (1994, in Russian). Translation: Russ. Acad. Sci. Sb. Math. **83**(1), 237–281 (1995)

179. A.V. Rybkin, On A-integrability of the spectral shift function of unitary operators arising in the Lax-Phillips scattering theory. Duke Math. J. **83**(3), 683–699 (1996)

180. L.A. Sahnovič, Dissipative operators with absolutely continuous spectrum. Trudy Moskov. Mat. Obšč. **19**, 211–270 (1968, in Russian)

181. J.T. Schwartz, *Nonlinear Functional Analysis* (Gordon and Breach Science Publishers, New York, 1969)

182. V.S. Shulman, Some remarks on Fuglede-Weiss theorem. Bull. Lond. Math. Soc. **28**, 385–392 (1996)

183. B. Simon, *Trace Ideals and Their Applications*. Mathematical Surveys and Monographs, 2nd edn., vol. 120 (American Mathematical Society, Providence, 2005)

184. A. Skripka, On properties of the ξ-function in semi-finite von Neumann algebras. Integr. Equ. Oper. Theory **62**(2), 247–267 (2008)

185. A. Skripka, Trace inequalities and spectral shift. Oper. Matrices **3**(2), 241–260 (2009)

186. A. Skripka, Higher order spectral shift, II. Unbounded case. Indiana Univ. Math. J. **59**(2), 691–706 (2010)

187. A. Skripka, Multiple operator integrals and spectral shift. Illinois J. Math. **55**(1), 305–324 (2011)

188. A. Skripka, Taylor approximations of operator functions, in *Operator Theory in Harmonic and Non-commutative Analysis*. Operator Theory Advances and Applications, vol. 240 (Birkhäuser, Basel, 2014), pp. 243–256

189. A. Skripka, Asymptotic expansions for trace functionals. J. Funct. Anal. **266**(5), 2845–2866 (2014)

190. A. Skripka, Trace formulas for multivariate operator functions. Integr. Equ. Oper. Theory **81**(4), 559–580 (2015)

191. A. Skripka, Estimates and trace formulas for unitary and resolvent comparable perturbations. Adv. Math. **311**, 481–509 (2017)

192. A. Skripka, On positivity of spectral shift functions. Linear Algebra Appl. **523**, 118–130 (2017)

193. A. Skripka, Taylor asymptotics of spectral action functionals. J. Oper. Theory **80**(1), 113–124 (2018)

194. A. Skripka, M. Zinchenko, Stability and uniqueness properties of Taylor approximations of matrix functions. Linear Algebra Appl. **582**, 218–236 (2019)

195. A. Skripka, M. Zinchenko, On uniqueness of higher order spectral shift functions. Stud. Math. (2019). https://doi.org//10.4064/sm181007-1-1

196. M.Z. Solomjak, V.V. Sten'kin, A certain class of multiple operator Stieltjes integrals, in *Linear Operators and Operator Equations*. Problems in Mathematical Analysis, no. 2 (Izdat Leningrad University, Leningrad, 1969, in Russian), pp. 122–134. Translation: Linear Operators and Operator Equations. Problems in Mathematical Analysis (Consultants Bureau, New York, 1971), pp. 99–108

197. E.M. Stein, *Singular Integrals and Differentiability Properties of Functions*. Princeton Mathematical Series, no. 30 (Princeton University Press, Princeton, 1970)

198. E.M. Stein, G. Weiss, *Introduction to Fourier Analysis on Euclidean Spaces*. Princeton Mathematical Series, no. 32 (Princeton University Press, Princeton, 1971)

199. V.V. Sten'kin, Multiple operator integrals. Izv. Vysš. Učebn. Zaved. Mat. **179**(4), 102–115 (1977, in Russian)

200. F. Sukochev, Hölder inequality for symmetric operator spaces and trace property of K-cycles. Bull. Lond. Math. Soc. **48**(4), 637–647 (2016)

201. K. Sundaresan, Smooth Banach spaces. Math. Ann. **173**, 191–199 (1967)

202. M. Takesaki, *Theory of Operator Algebras, I* (Springer, New York, 1979)

203. I.G. Todorov, L. Turowska, *Schur and Operator Multipliers*. Banach Algebras 2009 (Banach Center Publications 91, Institute of Mathematics, Polish Academy of Sciences, Warsaw, 2010), pp. 385–410

204. N. Tomczak-Jaegermann, On the Differentiability of the Norm in Trace Classes S_p. Séminaire Maurey-Schwartz 1974–1975: Espaces L^p, applications radonifiantes et géométrie des espaces de Banach, Exp. No. XXII (Centre Math., École Polytech., Paris, 1975), 9 pp.

205. A. Tomskova, Representation of operator Taylor remainders via multilinear Schur multipliers, preprint

206. M. Terp, L_p-spaces Associated with von Neumann Algebras. Notes, Math. Institute (Copenhagen University, Copenhagen, 1981)

207. H. Triebel, *Theory of Function Spaces*. Mathematik und ihre Anwendungen in Physik und Technik, vol. 38 (Akademische Verlagsgesellschaft Geest & Portig K.-G., Leipzig, 1983)

208. W. van Ackooij, B. de Pagter, F.A. Sukochev, Domains of infinitesimal generators of automorphism flows. J. Funct. Anal. **218**(2), 409–424 (2005)

209. A. van Daele, *Continuous Crossed Products and Type III von Neumann Algebras*. London Mathematical Society Lecture Note Series, vol. 31 (Cambridge University Press, Cambridge, 1978)

210. J. Voigt, On the convex compactness property for the strong operator topology. Note Mat. **12**, 259–269 (1992). Dedicated to the memory of Professor Gottfried Köthe

211. G. Weiss, The Fuglede commutativity theorem modulo the Hilbert-Schmidt class and generating functions for matrix operators, I. Trans. Am. Math. Soc. **246**, 193–209 (1978)

212. H. Widom, When are differentiable functions differentiable?, in *Linear and Complex Analysis Problem Book, 199 Research Problems*. Lecture Notes in Mathematics, vol. 1043 (Springer, Berlin, 1983), pp. 184–188

213. J.P. Williams, Derivation ranges: open problems, in *Topics on Modern Operator Theory*. Operator Theory: Advances and Applications, vol. 2 (Birkhäuser, Basel, 1981), pp. 319–329
214. D.R. Yafaev, *Mathematical Scattering Theory. General Theory*. Translations of Mathematical Monographs, vol. 105 (American Mathematical Society, Providence, 1992)
215. D.R. Yafaev, A trace formula for the Dirac operator. Bull. Lond. Math. Soc. **37**(6), 908–918 (2005)

Index

© Springer Nature Switzerland AG 2019
A. Skripka, A. Tomskova, *Multilinear Operator Integrals*,
Lecture Notes in Mathematics 2250, https://doi.org/10.1007/978-3-030-32406-3

LECTURE NOTES IN MATHEMATICS  Springer

Editors in Chief: J.-M. Morel, B. Teissier;

Editorial Policy

1. Lecture Notes aim to report new developments in all areas of mathematics and their applications – quickly, informally and at a high level. Mathematical texts analysing new developments in modelling and numerical simulation are welcome.

   Manuscripts should be reasonably self-contained and rounded off. Thus they may, and often will, present not only results of the author but also related work by other people. They may be based on specialised lecture courses. Furthermore, the manuscripts should provide sufficient motivation, examples and applications. This clearly distinguishes Lecture Notes from journal articles or technical reports which normally are very concise. Articles intended for a journal but too long to be accepted by most journals, usually do not have this "lecture notes" character. For similar reasons it is unusual for doctoral theses to be accepted for the Lecture Notes series, though habilitation theses may be appropriate.

2. Besides monographs, multi-author manuscripts resulting from SUMMER SCHOOLS or similar INTENSIVE COURSES are welcome, provided their objective was held to present an active mathematical topic to an audience at the beginning or intermediate graduate level (a list of participants should be provided).

   The resulting manuscript should not be just a collection of course notes, but should require advance planning and coordination among the main lecturers. The subject matter should dictate the structure of the book. This structure should be motivated and explained in a scientific introduction, and the notation, references, index and formulation of results should be, if possible, unified by the editors. Each contribution should have an abstract and an introduction referring to the other contributions. In other words, more preparatory work must go into a multi-authored volume than simply assembling a disparate collection of papers, communicated at the event.

3. Manuscripts should be submitted either online at www.editorialmanager.com/lnm to Springer's mathematics editorial in Heidelberg, or electronically to one of the series editors. Authors should be aware that incomplete or insufficiently close-to-final manuscripts almost always result in longer refereeing times and nevertheless unclear referees' recommendations, making further refereeing of a final draft necessary. The strict minimum amount of material that will be considered should include a detailed outline describing the planned contents of each chapter, a bibliography and several sample chapters. Parallel submission of a manuscript to another publisher while under consideration for LNM is not acceptable and can lead to rejection.

4. In general, **monographs** will be sent out to at least 2 external referees for evaluation.

   A final decision to publish can be made only on the basis of the complete manuscript, however a refereeing process leading to a preliminary decision can be based on a pre-final or incomplete manuscript.

   Volume Editors of **multi-author works** are expected to arrange for the refereeing, to the usual scientific standards, of the individual contributions. If the resulting reports can be forwarded to the LNM Editorial Board, this is very helpful. If no reports are forwarded or if other questions remain unclear in respect of homogeneity etc, the series editors may wish to consult external referees for an overall evaluation of the volume.

5. Manuscripts should in general be submitted in English. Final manuscripts should contain at least 100 pages of mathematical text and should always include

   - a table of contents;
   - an informative introduction, with adequate motivation and perhaps some historical remarks: it should be accessible to a reader not intimately familiar with the topic treated;
   - a subject index: as a rule this is genuinely helpful for the reader.
   - For evaluation purposes, manuscripts should be submitted as pdf files.

6. Careful preparation of the manuscripts will help keep production time short besides ensuring satisfactory appearance of the finished book in print and online. After acceptance of the manuscript authors will be asked to prepare the final LaTeX source files (see LaTeX templates online: https://www.springer.com/gb/authors-editors/book-authors-editors/manuscriptpreparation/5636) plus the corresponding pdf- or zipped ps-file. The LaTeX source files are essential for producing the full-text online version of the book, see http://link.springer.com/bookseries/304 for the existing online volumes of LNM). The technical production of a Lecture Notes volume takes approximately 12 weeks. Additional instructions, if necessary, are available on request from lnm@springer.com.

7. Authors receive a total of 30 free copies of their volume and free access to their book on SpringerLink, but no royalties. They are entitled to a discount of 33.3 % on the price of Springer books purchased for their personal use, if ordering directly from Springer.

8. Commitment to publish is made by a *Publishing Agreement*; contributing authors of multiauthor books are requested to sign a *Consent to Publish form*. Springer-Verlag registers the copyright for each volume. Authors are free to reuse material contained in their LNM volumes in later publications: a brief written (or e-mail) request for formal permission is sufficient.

Addresses:
Professor Jean-Michel Morel, CMLA, École Normale Supérieure de Cachan, France
E-mail: moreljeanmichel@gmail.com

Professor Bernard Teissier, Equipe Géométrie et Dynamique,
Institut de Mathématiques de Jussieu – Paris Rive Gauche, Paris, France
E-mail: bernard.teissier@imj-prg.fr

Springer: Ute McCrory, Mathematics, Heidelberg, Germany,
E-mail: lnm@springer.com

Printed in the United States
By Bookmasters